ANACONDA MONTANA

COPPER SMELTING BOOM TOWN
ON THE WESTERN FRONTIER

by

Patrick F. Morris

History is the raw material of self-definition.
By exploring the past we locate ourselves in the present.

ANACONDA MONTANA
Copper Smelting Boom Town on the Western Frontier
by Patrick F. Morris

Published by:
Swann Publishing
5005 Baltan Rd.
Bethesda, MD. 20816

Suggested Library of Congress Cataloging in Publication Data

Morris, Patrick F.
    Anaconda Montana, Copper Smelting Boom Town on the
Western Frontier.

    Includes Bibliography and Index
    ISBN 0-9657209-2-6 : $16.95
    1. Social Conditions--Anaconda--History. I. Title
    2. Gold and Silver mining--West--Montana--History.
    3. Copper industry--Butte, Anaconda--History.
    4. Politics, Labor Unions--West--Montana--History.

0978.78687   96-93090

DEDICATION

This book is dedicated to everyone who ever called Anaconda home, and to my grandfather, John James "Jack" Morris, (born Morrisroe in County Mayo, Ireland) who participated in or was witness to most of the events related herein. He helped build the foundation of Anaconda's first smelter, built and ran one of its first boarding houses and saloons and was a member of the first city council. He was the first of his large family of Irish immigrants to settle in Anaconda and Butte and later in other parts of the West. He died in 1923 and is buried in Anaconda's Hill Cemetery. Since I owe my own existence and birth in Anaconda to him, it is no exaggeration to say that if weren't for him this book would not have been written.

CONTENTS

FOREWORD

This is the story of the founding of a small western town in the last decades of the nineteenth century and its life during the first two decades of the twentieth century. No city, town or individual lives in complete isolation. Each can be understood only within the context of its greater universe. I, therefore, have extended this narrative beyond the listing of facts and events within the immediate confines of the town itself and have tried to identify and describe the multiple connections, both physical and historical, which influenced and molded it.

In pursuing this endeavor I am indebted to all those who preceded me in recording Anaconda's past. I particularly want to mention: the Deer Lodge County History Group which compiled that wonderful collection of memories of Anaconda *In the Shadow of Mt. Haggin*, as well as the works of Matt Kelly, Bob Vine, Mary Dolan, Ruth Meidl, George Wellcome and Hugh Campbell all of which I consulted frequently. And I want to commend the wonderful spirit of community which created and sustains that repository of the town's historical consciousness The Anaconda Deer Lodge County Historical Society.

I am especially grateful to Alice Finnegan, one of its founders, who encouraged me during the initial stages of this project and generously shared with me her own research, provided me information and leads and reviewed some of my early chapters; to Bob Vine and Jerry Hansen of the Historical Society for giving me access to their records and picture files and for their continuing interest and cooperation throughout; to Natalie Sliepcevich and Marian Geil of the Hearst Free Library for their friendly assistance and to Ellen Crain of the Butte Silver Bow Archives for helping me with my research and for reviewing early chapters. Finally, I want to thank the First Montana Heritage Park & Partners for endorsing this book as a contribution to its efforts to establish the Anaconda-Butte Heritage Corridor as a Labor History Landmark District. All pictures are courtesy of The Anaconda Deer Lodge County Historical Society. The cover was designed by my son Robert Kevin Morris, and Paul Irey.

ANACONDA AND VICINITY
At turn of the century

1. *"Mike, We've Got It!"*

She wasn't pretty but she wasn't plain either. I think what caught people's attention and made her stand out in their minds was her fierce pride. When she was at her best that quality made her sparkle with brilliance. And it also gave her great dignity during the rough times in her life.

It may seem odd to describe ones home town in such terms, but as I observe her now slipping quietly out of one century into another she seems to me like a dignified old lady that I've known all my life who has many tales to tell if we'll just stop awhile and listen. She has seen much and suffered some, but she has had her share of joy and excitement as well. And she bares her one hundred and fourteen years with great grace and dignity.

While much of her story has been told in bits and pieces before, those accounts were usually about individuals and events told from someone else's perspective or tagged onto a yarn about something else or someplace else. Now she wants to tell what happened from her own point of view. And though she may from time to time wander outside her immediate environs in the telling, you can be sure that it all has a bearing on her life.

She was only seventeen years old when the century began and was bursting with an energy and confidence that reflected the spirit of the nation as a whole. Indeed, her growth and health typified the promise of America. But before we get into that, she wants you to know a little about how she came to be and how she got her name and maybe even where that fierce pride of hers came from.

It would be stretching things a bit to say that she was conceived in the Civil War, but that's where her name came from and she wants everyone to know that she's a "one hundred percent American creation" and likes to think of herself as one of the positive things that came out of that great conflict. Of course, she knows that a lot of other forces in the world at that time also had a bearing on her birth, but prefers to stress her American roots and American upbringing.

The two men who had most to do with providing her name were an obscure Irish immigrant named Thomas Hickey and the famous Horace Greeley, who, besides saying "Go west, young man" also wrote in the New York Herald newspaper (of which he was editor) predicting that the Union armies would "encircle Lee's forces and crush them like a giant anaconda." Young Hickey, who was a Union soldier in the Civil War, recalled this powerful image and also heeded Greeley's advice about going west. He carried the name west with him after the war and in 1875 gave it to one of his numerous mining claims on a hill that was later to be called Butte.[1]

Butte was already on its way to becoming a giant roaring mining camp that would also come to be known as the "richest hill on earth". You could say that Butte was Anaconda's older brother, and this is more than just a figure of speech. There was always a strong feeling of kinship between the two towns despite occasional family quarrels and ongoing rivalries. And like any family, their stories are so tangled that in order to make sense of one you've got to tell about the other, too. In the past when this has been done Anaconda as a town has gotten lost in the telling, and that is what we're now setting out to remedy.

In 1880, Marcus Daly, another Irish immigrant, working for the Walker brothers of Salt Lake City, bought Hickey's one-third interest in the Anaconda mine for $10,000 and paid an additional $20,000 to Hickey's partner Charles X. Larabie for the other two-thirds. Thus began the saga of first a mine, then a company, and finally a town, all with the name, "Anaconda".[2]

Daly began immediate development work on the mine and confirmed his belief that it was rich in silver ore with strong traces of copper. Because the Walker brothers did not want to make the large investment necessary to develop the mine, Daly went to an old friend from his silver mining days in Nevada and Utah, George Hearst. Hearst, on Daly's recommendation, had bought the Ontario silver mine in Utah, for a short time known as the world's richest, and had become wealthy from it. After examining the Anaconda property, Hearst recommended that the syndicate, in which he was a partner with James Ben Ali Haggin and Lloyd Tevis, purchase the mine and let Daly develop it. The syndicate bought the mine from Daly for what he had paid for it and gave him one quarter ownership and an ample drawing account for developing and equipping it. In addition to Daly's share, Hearst owned thirty-nine per cent, Haggin twenty-six and Tevis ten.[3]

Work on the mine began in June 1881. The ore was crushed in a stamp mill rented from another mineowner, W. A. Clark, and within months the mine was producing oxidized silver ore running at thirty ounces a ton. But from the start, Daly had also been interested in the mine's copper potential. Late in 1882, at a depth of 300 feet, as the silver vein in the mine seemed to be playing out, Daly's miners ran into a "new material." When the dust cleared from a blast, Daly picked up a piece of ore that sparkled from its high metal content. It was copper glance and Daly recognized it immediately. He looked at his foreman, Mike Carroll and exclaimed, "Mike, we've got it."[4]

This was it. They had just discovered the largest deposit of copper sulphide the world had ever known. It would result not only in large-scale mining of copper out of Butte hill, but would eventually make the United States the world's leading copper producer. If there was any single moment in the long sequence of events that determined the founding of the town of Anaconda, this was it. It was a magic moment that set the stage for all that was to follow.[5]

But it was neither copper nor silver which set off the original mad scramble for mineral wealth in the West and led up to this moment and the founding of the town. It was gold. Gold, that infernal, shiny, yellow metal, held the same hypnotic fascination for the citizens of the sprawling, boisterous, young United States stretched along the east coast and into the hinterland of North America that it did for Columbus and the Spanish who conquered and ruled large portions of the Western Hemisphere before the arrival of the Dutch, English and French. In 1849, when it was discovered of easy access and in unbelievably large quantities on the banks of Sutter's Creek in far-off California, it set in motion a tide of migration and settlement across the wide expanse of the continent that rivaled that of the Germanic tribes in Europe at the fall of Rome. Once gold fever took hold of the nation, it wasn't cooled by the eventual domination and depletion of the California bonanza, which yielded over $500 million in its first ten years. The search was on all over the West. Those who didn't make it in California or who came too late moved on to wherever there was even a rumor of a new strike.

Ten years after the California discovery, a new find was made in Colorado in the foothills of the Rockies near Pikes Peak and wagon trains with the motto "Pikes Peak or Bust" scrawled on them pushed through Indian lands to get in on the riches. New towns of Golden, Boulder, Denver City and Colorado City sprang up almost overnight. "Cities of gold, lust and death, sprang up and then died away again," wrote a distinguished visitor from England, Robert Louis Stevenson.[6]

In the same year as the Colorado discovery, silver mixed with gold was found on the eastern slope of the Sierra Nevada, near Lake Tahoe by two immigrant Irishmen, Peter O'Reily and Pat McLaughlin. Their find, the Comstock lode, would eventually become known as the richest silver vein in the world. Within a year, ten thousand men were tunneling into the mountain sides and the roaring mining camps of Virginia City, Aurora and Gold Hill came to life on the arid slopes of Mt. Davidson. "Here in these uncouth places pig-tailed Chinese worked side by side with border ruffians and broken men from Europe talking together in mixed dialects and mostly oaths," wrote a disapproving Stevenson. Later silver finds in Colorado, New Mexico and Arizona sprouted yet more new towns with such descriptive names as Leadville, Silver City and Tombstone.[7]

In 1860, gold was discovered on the Nez Perces reservation in Idaho, attracting a wave of prospectors from California, Oregon and Nevada and before long new towns of Pierce, Orofino, and Lewiston were founded. Then more discoveries created Elk City and Florence, followed by still others in the Boise Basin which gave birth to Placerville, Centerville and Idaho City. The trading post and stage stop of Boise became Boise City. As early as January 1862, over $3 million in gold had been shipped out of these camps. But as soon as the gold ran out the towns were abandoned as fast as they were populated. Silver strikes in northern Utah became the next stop for many of the fortune-seekers, and others rushed into Montana when gold was found there.[8]

At the same time, lonely prospectors were exploring every stream, gulch and valley in the West looking for placer gold, that is loose nuggets, flakes and specks, the kind of gold that the forces of nature have separated from their source and washed down into alluvial sands on river banks and flood plains, the kind that was found on Sutter's Creek in California. Whenever word got out of even a small find, the rush was on. They were called "stampedes".

The Deer Lodge Valley, where infant Anaconda was soon to take up residence, and its surrounding hills, were to figure prominently in that exploration. The first recorded discovery of gold in Montana occurred at Gold Creek, first called American Fork, at the mouth of the valley, in 1852. It was worked intermittently from 1858 through 1860. But it was the find on Grasshopper Creek at Bannack, near present day Dillon, in 1862, that caused the first "gold rush" into what was then part of the Washington Territory. Within a year, Bannack had more miners and mines than it did paying claims. A fortuitous discovery to the east in Alder Gulch the following year offered the many who hadn't made it at Bannack another chance.[9]

With the outbreak of the Civil War, many a refugee from its slaughter and plunder, both Northerner and Southerner, became gold seekers. So many Southerners flocked to Alder Gulch that the first town on the site was called Varina after the wife of Jefferson Davis, President of the Confederacy. It was later changed to Virginia City, like its predecessor in Nevada, as an artful way to placate federal officials still sensitive over the recent conflict. Internal Revenue Collector N.P. Langford wrote from Virginia city in 1866: "Four-fifths of our citizens were openly declared Secessionists."[10]

Bannack's residential district was called "Yankee Flat." Confederate Gulch near Helena became one of the most sensational, if short lived, finds in the territory. A contemporary chronicler said that Montana in the late sixties was full of gold, whiskey, Indians and fugitive rebels. Another story had it that the left wing of General Sterling Price's Confederate Army, which was never mustered out, became the Montana Democratic Party.[11]

Virginia City had scarcely been established when word came of another gold strike at Last Chance Gulch, the site of present day Helena, and a new rush ensued. These three finds within such a short time of each other put Montana on the mining map. As in the case of other such mining camps, as soon as the gold was depleted, the towns were abandoned. Of the three, only Helena survives as an active community today.[12]

Although there is evidence that mining took place on what was to become Butte hill as early as 1858, it wasn't until the summer of 1863 when William Allison, Bud Parker, Pete McMahon and others found placer gold in quantity that other prospectors began moving into the area, some from Bannack, others from Colorado and Nevada to work the placers on Silver Bow Creek. By the end of that year there were more than a thousand men prospecting and living along the banks of the creek, named by McMahon for the artful silver bows it cut as it meandered westward into the Deer Lodge Valley. The next year Allison and G.O. Humphreys staked a claim and began digging on Butte hill itself in what can be called the beginning of quartz mining in the area.[13]

Isaac Marcosson, in his official history of the Anaconda Copper Mining company, says: "In 1865, Butte was a rip-roaring camp, tough, lawless, with nearly every man carrying a brace of pistols and a bowie knife stuck in one of his boots. Gold and silver were wrested in quantity from the placers. Then mining history began to repeat itself. The placers were worked out." By the early seventies, Silver Bow City, seven miles west of Butte which had boasted a population of over 1000 and was the first seat of Deer Lodge county, was a ghost town. Nevertheless, in the preceding years from 1862 to 1868, gold and silver extracted from southwest Montana was valued at over $90 million. Only California produced more gold during the period.[14]

Meanwhile on Butte hill quartz claims had been staked, shafts sunk and tunnels bored. It was the beginning of a more serious long-term effort to follow silver-rich veins into the mountainside. The townsite of Butte, laid out in 1867, had only a few hundred miners after the placers along Silver Bow Creek were depleted. It wasn't until 1874 when William L. Farlin began to work the Travona, a claim he had staked out a few years before, and was rewarded with silver in commercial quantities, that the town began to grow. Soon primitive mills and smelters dotted the landscape and silver production began to rise. By 1875 the population had increased to 4000.

While Butte was known throughout the seventies as a silver camp, copper ore was also being mined and concentrated, and from one property, the Parrot mine, half the ore shipped was copper. Even earlier, in 1868, Joseph Ramsdell, Billy Parks, Dennis Leary and Charles Porter constructed a primitive copper smelter at the Parrot, which was soon abandoned because the ore proved too complex. It is also interesting to note that W.A. Clark, later to become Marcus Daly's chief rival, who took over the Travona mine from Farlin and owned other claims on the hill, had also shipped copper to Utah in 1873 and 1874, three years before Daly arrived in Butte. By August 1879, Clark was operating a large smelter on the south side of Silver Bow creek with his partners in the Colorado and Montana Company.[15]

But it was silver that attracted Daly and his employers, the Walker brothers, to Butte in 1876, where they bought the Alice mine. Daly, who operated the mine for them, was one of the many immigrant Irish who had followed gold west to California, then Nevada and Utah and finally Montana.[16]

Born in Ireland near the village of Ballyjamesduff, County Cavan in 1841, Daly was part of the great exodus from Ireland to the United States that followed the potato famine of 1845-48. He arrived in New York city at age 15 in 1856, ready to take any work that was offered. After working on the docks in Brooklyn, he managed to wangle passage on a ship to Panama in 1861, and thence across the isthmus and on to San Francisco where his married sister Ann O'Farrel lived. Fascinated by the tales of gold and quick fortunes, he wasted little time in making his way to the diggings in northern California, joining scores of his countrymen similarly enticed.

Although the time of quick fortunes from California gold was over, Daly gained experience in both placer and hardrock mining in Calaveras County, before moving on to Nevada. There he became a foreman at the "Grand Bonanza" Comstock mine, under John Mackay, who would later accumulate one of Nevada's largest fortunes in partnership with three other Irishmen, Fair, Flood and O'Brien. While living in Virginia City, Nevada, Daly made a wide range of acquaintances which served him well in later years. It was there that he met George Hearst, a mining entrepreneur, working with the Haggin-Tevis syndicate out of San Francisco. He also had a passing acquaintance with Mark Twain who was then a young reporter on the *Territorial Enterprise* newspaper. It was there too that Daly not only perfected his mining skills but developed an uncanny instinct for sensing the presence or absence of mineral wealth in a particular location. It would take all of this accumulated experience and his native shrewdness to transform his spectacular find at Butte into a paying enterprise. But it was pure coincidence that he chose to build his smelter in the same valley that had earlier been the site of the territory's first gold strike.[17]

It was in 1857 when the first placers were worked on Gold Creek, where the Deer Lodge River becomes the Hells Gate. By 1858, a small community of six cabins and various tepees was inhabited, but later abandoned, by the Stuart brothers (James and Granville) Rezin Anderson, Sterney Blake, Bud McAdow and others. In 1860, Henry Thomas sank a 30-foot shaft in the same drift. The Stuarts returned in May 1862 and set up the first sluices in Montana near the head of Gold Creek. History notes that there were 45 people there for a July 4 celebration in 1862. A letter written by the Stuarts to their brother Thomas in Colorado, advising him to join them, contributed to a stampede to Montana, which included John M. Bozeman, who in 1863 blazed a new road from the Oregon trail to the Montana gold camps. Also attracted to the Gold Creek diggings was young, Kentucky-born Samuel T. Hauser, destined to figure prominently in Montana history. [18]

These weren't the first white men the area had seen. Although the Lewis and Clark expedition of 1810 skirted Deer Lodge Valley in its trek west, soon thereafter white trappers and mountain men began to rendezvous there with Indians and fur traders from the British owned Hudson Bay company and John Jacob Astor's American Fur Company, who passed through once or twice a year. History records that Flathead (Salish) Indians guided brigades from the Hudson Bay Company through Deer Lodge, Big Hole and Beaverhead valleys in 1818. On July 29, 1825 Peter Ogden, working for Hudson Bay Company's Snake Country Expedition, moved north from the Big Hole "over very hilly country" to camp on a large stream that "discharges into the Columbia". Moving up Warm Springs Creek Ogden found "Mountains on all sides high and lofty. Goats and red deer in abundance. Beaver very scarce." Others who knew the valley were Finan McDonald, Alexander Ross, Jim Bridger, Tom Fitzpatrick and Jedediah Smith.[19]

In 1840, the Belgian Jesuit, Pierre Jean de Smet, passed through and celebrated mass at a large mound on the valley floor which spouted steam and hot water called by the Indians "The Lodge of the White Tailed Deer." It was from this mound that the valley and the river flowing close by came to be called Deer Lodge. De Smet had been invited by the Flathead Indians to establish the St. Mary's mission in the Bitter Root Valley, the next large valley to the west. He was joined there in 1845 by the Italian Jesuit, Antonio

Ravalli. Further north, another Jesuit mission was established by Father Adrian Hoecken in the Mission Valley in 1854.[20]

In 1853, Mexican War hero Isaac I. Stevens, who was commissioned by the US government to survey a railroad route from Lake Superior to the Puget Sound, left a survey party in the Bitter Root Valley, under twenty-five year old Lieutenant John Mullan, to explore and map the entire area, which included the Big Hole, Deer Lodge and Beaverhead valleys. Mullan had earlier that year found the pass across the continental divide west of Helena which now bears his name.[21]

It was 1857 when "Captain" Richard Grant and his son John moved several hundred head of cattle first into the Beaverhead and then into the Deer Lodge Valley. John settled where the Little Blackfoot River flows into the Deer Lodge. His father went further down river to Hell Gate Ronde (Missoula). In 1859, after having sold 400 head of cattle to be driven to California, John moved further up river where he built a two story house with glass windows and green shutters. Along with Louis Demers, Leon Quesnelle and Thomas LaVatta he became one of the original settlers in the little town of Cottonwood, which was incorporated two years later as Deer Lodge.[22]

In 1859, Lieutenant Mullan began constructing a military road from Fort Walla Walla on the Columbia River eastward toward Fort Benton on the Missouri. He worked his way north across the Snake to Coeur d'Alenes Lake, over the Bitter Root mountains through the Sohon pass, along the Clark Fork past Hell Gate and Gold Creek, and over the continental divide at Mullan pass. He reached Fort Benton on August 1, 1860. This road became a principal route for gold seekers from the east rushing to the new gold finds in Idaho.[23]

By 1862, however, word that Florence and Elk City were already played out diverted many to try their luck in Montana. It was at that time the streams and hills in the general vicinity of Daly's future town of Anaconda were explored for gold and silver. From 1862 to 1882, every gulch and stream running into Deer Lodge Valley and beyond was combed over and with surprising frequency new discoveries were made, east, west, north and south.

On July 12, 1862, Mortimer Lott found color at Wisdom near the Big Hole River about 50 miles southeast of present day Anaconda. Then, another 50 miles further south, on July 28, John White, William Eads and the Morley brothers filed claim at Grasshopper Creek and set off the Bannack gold stampede. In 1864, Mortimer Lott made another find at French Gulch, only 20 miles from the future Anaconda townsite. This find set off a small stampede and produced over $5 million in gold over the next five years.[24]

Similar activity was going on to the north and west of the future town, as well. On June 14, 1864, Henry Horton filed a location notice for a discovery of silver ore on Hope Hill, Deer Lodge county, near the present town of Philipsburg, thirty miles west of Anaconda. In subsequent years, the Cordova, Hope, Home Sweet Home and other Hope Hill claims opened up rich bodies of silver ore. The collection of miners' shacks known as Camp Creek became Philipsburg in 1867, named after Philip Deidesheimer, superintendent of a silver company, and the plat of the Philipsburg townsite was entered on the books of Deer Lodge county April 26, 1876. There were at that time two other small settlements near mines at the heads of gulches above the town named Hasmark and Stumptown and groups of shacks and houses at nearby Kirkville and Rosalind.

In 1875, Eli D. Holland shot a deer near the peak of Granite Mountain, ten miles east of Philipsburg. As he was recovering it, he noticed an outcropping that turned out to be silver and another mine was started. But it wasn't until 1882, when Charles McClure with investment capital from a mining syndicate out of St. Louis, Missouri, blasted open a major silver vein that a real stampede to the area began. Within a few years, Granite was a growing, prosperous town attracting as many newcomers as Daly's town 28 miles east on Warm Springs Creek.[25]

Over $20 million was taken out of Granite before the repeal of the Sherman Silver Act and the panic of 1893. When that occurred, it is said that three thousand people left Granite within twenty-four hours and it became a ghost town. However, after a few years the two great silver properties of Granite and BiMetalic were consolidated under a single company and production resumed. A dam was built on Flint Creek near Georgetown, 12 miles east, to bring electricity to Granite and the Georgetown flat became a lake. From 1898 through 1904, the mines produced $1 million a year.[26]

In 1867, Alexander Aiken, John E. Pearson and Jonas Stough discovered, just 15 miles west of the future Anaconda, what came to be called the Cable quartz load on Cable Mountain. The name commemorated the laying of the second trans-Atlantic cable. The mine was operated with indifferent success until 1880 when extremely rich ore was found. One 500 foot piece of ground yielded $6-,500,000 in gold. In 1889, it produced the largest gold nugget ever found which was later purchased by W.A. Clark for $10,000. Soon - after the Cable discoveries, finds were made at Southern Cross and Georgetown on upper Flint Creek and small communities grew up near those mines.

Georgetown, named after George Cameron, brother of Salton Cameron mine operator at both Cable and Southern Cross, was originally a small lumber camp. In 1876, Tom Stuart and O.B. Whitford built a five-stamp mill powered by a water wheel near the Georgetown diggings. In the following years, more gold was discovered and new mines opened in this area. The Red Lion and Gold Coin mines opened in 1896 and have been sporadically operated on and off up to the present day, depending upon the price of gold.[27]

Some prospectors from the East, after crossing the Mullan pass and reaching the Deer Lodge River abandoned their search for gold and settled along its banks to raise cattle and farm. One reason for this change of heart may be garnered from an entry in the diary of James Morley on July 7, 1862, at his first sight of the valley. His sentiment has as much resonance today as when it was written: "Was there in this world ever a country half so pretty as was this mountain land before civilization ruined it?" Morley wrote this only three weeks before his historic find at Grasshopper Creek. He and his brother had originally intended to go to gold diggings further west in Idaho.[28]

But not all of the gold seekers were as lucky as the Morley brothers, so were forced to turn to other pursuits, some of which would prove to be as lucrative as mining. The grasses of Deer Lodge Valley were ideal for grazing cattle and raising horses. As the number of gold and silver mines increased the demand for both grew. Johnny Grant, Thomas Lavata, Alejo Barasta, and Joseph Hill were already raising cattle down river from Deer Lodge in 1861 and Bob Dempsey, Dave Courtway and Thomas Irvine established ranches just up river from the town near Race Track Creek. Later the Prouse family bought the Dempsey place and the Quinlan family settled nearby. Others who began raising cattle in that part of the valley during those years were Frank Mason, Louis DeMar, Leon Cannel and L.R. Maillett.[29]

By 1865, when David D. Walker began raising cattle on Cottonwood Creek, the town of Deer Lodge had 125 cabins and three steam saw mills. The Stuart brothers and Walter Dance operated one which produced 15,000 feet of lumber a day. Johnny Grant operated a grist mill and had a thresher to support the mill. By 1869, a courthouse had been erected and plans were underway to build a territorial penitentiary there. That was also the year that W.A. Clark in partnership with R.W. Donnel and S.E. Larabie began a wholesale merchandising and banking business there. Clark would later establish a branch of that bank in Butte under the name of W.A. Clark and Brother, Bankers.[30]

David Walker, born in Iowa in 1843, continued raising cattle in the valley, but moved to Anaconda in 1883, when the town was founded, bought one of the first lots and opened a butcher shop in partnership with Nick Beilenberg, a recent immigrant from Germany. Walker would become one of the early mayors of Anaconda.[31]

The gold strikes at Bannack, Virginia City, Helena and Silver Bow created a growing market for meat and horses. In many cases, beef was the only thing available to eat in the mining camps. A gulch near Silver Bow was given the name Beef Straight Gulch because one season the miners working that stretch of creek had run out of flour and lived on nothing but beef for three months.

Cottonwood was well situated to supply those camps and was joined to the southeast by a rough, winding road that made its way along the foothills to Silver Bow. Joseph Moog, a German prospector, was hired by the Silver Bow-Cottonwood Coach Line to establish a stage station on that road which they named Stuart. Moog, who had moved from Golden Colorado to French Gulch before settling in Stuart, stayed there and raised a family. His daughter Mary Montana Moog and Josie Beecher were the first graduates of Butte High School.[32]

Conrad Kohrs, a German immigrant from Schleswig-Holstein, who first came into the valley over that road in 1862, was a prospector who became a cattleman as he passed through the valley. He observed, as his small party made its way down Silver Bow creek, "It was a beautiful stream, the water clear and sparkling and alive with the finest trout, and the same was true of every small stream we crossed. The valley was full of antelope and herds of fat cattle belonging to the mountaineers who lived there." He had run out of money so was happy to take a job as a butcher for $25 a month. The man who hired him, Hank Crawford, wanted him to buy cattle, herd them over the continental divide to Bannack and butcher and sell the meat there. Within two years, Kohrs had a herd of over 400 head of cattle grazing on the Dodge ranch near Race Track.[33]

He brought his half brothers, John, Charlie and Nick Beilenberg from Germany and set them up in butcher shops in Helena, Deer Lodge and Blackfoot. Nick later opened a butchershop in Butte and joined David Walker and opened another in Anaconda. In 1866, Kohrs bought out Johnny Grant for $19,200 after Grant was busted by federal authorities for illegally selling liquor to the Indians. Kohrs was on his way to becoming the largest and most successful cattle dealer in the Territory. He was not only providing his own butcher shops, but was selling meat wholesale and cattle on the hoof. In subsequent years, he expanded his cattle empire east to the Sun River and began providing cattle to the eastern markets through Chicago. He always insisted that cattle raising in Montana was started in the Deer Lodge Valley before the cattle drives from Texas brought stock to the ranges of eastern Montana. [34]

DEER LODGE VALLEY AND EARLY GOLD FINDS IN SOUTHWESTERN MONTANA

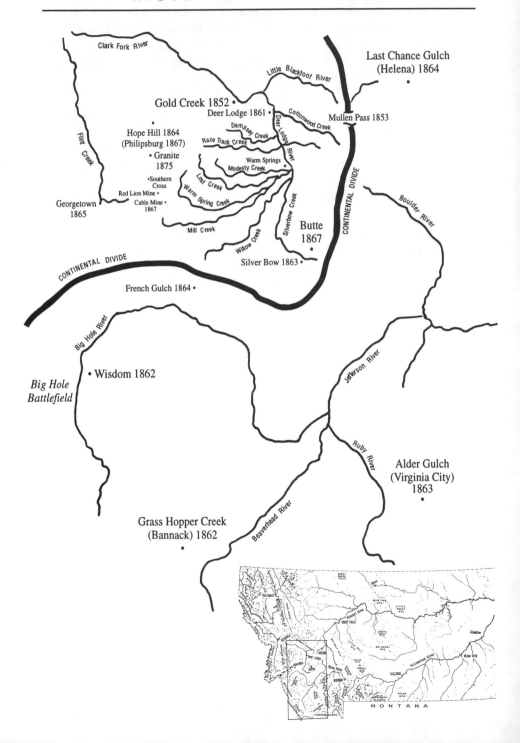

Clark Fork River

Little Blackfoot River

Last Chance Gulch (Helena) 1864

Gold Creek 1852 •

Deer Lodge 1861 •

Cottonwood Creek

Mullen Pass 1853

Deer Lodge River

Hope Hill 1864 (Philipsburg 1867)

Dempsey Creek

Race Track Creek

Flint Creek

• Granite 1875

Warm Springs

Modesty Creek

•Southern Cross

Red Lion Mine •

Cable Mine • 1867

Lost Creek

Warm Spring Creek

CONTINENTAL DIVIDE

Boulder River

Georgetown 1865

Mill Creek

Willow Creek

Silverbow Creek

Butte 1867

Silver Bow 1863 •

CONTINENTAL DIVIDE

French Gulch 1864 •

Big Hole River

Jefferson River

Big Hole Battlefield

• Wisdom 1862

Ruby River

Alder Gulch (Virginia City) 1863

Grass Hopper Creek (Bannack) 1862

Beaverhead River

MONTANA

2. Heaven on Earth

It is clear then that upper Warm Springs Creek, where Dame Anaconda was to plant her roots, had been traversed, explored and prospected well before Marcus Daly decided to build his smelter there. In fact, twenty years earlier, about the time that Silver Bow Creek was attracting gold seekers, the first pioneers began arriving in Deer Lodge Valley to make their homes. While some of this settlement was a spillover from the gold rushes at Bannack, Virginia City, Silver Bow and French Gulch, the valley also became the refuge of a small band of dissident Mormons, who had formed a splinter group under an English convert to Mormonism, Joseph Morris.

Morris had migrated to Utah in 1853. He was convinced that God had spoken to him and selected him as a prophet. He challenged Brigham Young's leadership and precipitated a show-down with him. In June, 1862, a posse of 500 armed Mormons attacked the Morrisites at their settlement on the banks of the Weber River, near Ogden. The Morrisites, outnumbered in the territory by forty to one, resisted for three days and then surrendered after their leader and several of his followers were killed. Following this, the leaderless band left Utah and dispersed—some migrating to Idaho, Nevada and Montana, some returning to Iowa. In 1863, George Williams, a disciple of Morris, claimed to be the prophet Cainan and to have been ordained by God to assume leadership of the scattered flock.[1]

In that same year a number of Morrisite families settled in Deer Lodge Valley. One of those early settlers was William Price Thomas, a Welshman and confidant of the prophet Cainan, who located on Warm Springs Creek about three miles east of where Daly's future town would be founded. Thomas had four sons: John, Evan, Frank and Daniel, who later established homesteads of their own in the valley. Another settler was Ben Phillips who built a house near a series of hot springs at the southeastern end of the valley. This area was later named for Eli and George Gregson, who arrived in 1869, and developed the springs into a recreational area.[2]

In 1866, Morgan Evans, a Welshman, who became Daly's chief agent in acquiring the land for his new smelter and town, came into the valley with his family and built a log cabin near the Phillips ranch. Evans had trekked west with the Mormons to Corrine, Utah and had been a farmer until he heard of the gold rush at Alder Gulch and started a freight business between Salt Lake and Virginia City. On a trip from there to the town of Deer Lodge in 1865, he viewed the valley for the first time and was captivated by it. He returned to Virginia City, packed up his wife, mother, five children and all their belongings in two covered wagons and set out to make a new life for them. Evans, who prospered in the following years, had a 1000-acre ranch with 800 head of cattle, raised race horses, hay and grain and had his own grist mill. He fathered five more children, built an imposing house for his family further west on Mill Creek and named it Gwendale after his mother Mrs. Gwenilian Evans. He also established a post office by the same name, which she ran. Mrs. Evans had the distinction of being the first woman in Montana to be granted a homestead.[3]

While some of the Morrisites left the valley in 1866, others under William James, who recognized Williams as the prophet Cainan, moved in to replace them the following year. Then in 1871, a larger migration from Iowa, Nebraska, Utah and Idaho took place. It was June when a caravan of thirteen covered wagons came over the hump from Butte into the valley. That migration consisted of a number of Scandinavian families,(mostly Danes) the Eliasons, Jensens, the Jessens, Staffensons, Rasmussens, Jorgensons, Petersons, Johnsons, Hendricksons, Stroms, Orms and Hengalls, as well as English, Scotch and Welsh Morrisite believers such as Daniel James, the leader of the group, Fred J. Bliss, Angus Smith, A. M. Walker, J.R. Eardley, J. Howel, and M. Fifer.

The Morrisites established two separate congregations, called in the Mormon way, stakes, one in the upper valley near Lost Creek and the other closer to Deer Lodge near Race Track Creek. Meeting houses and schools were built at both locations. Total membership in the two stakes, including all members of the families, was about eighty people. They called their sect the Church of Jesus Christ of the Saints Most High and they referred to what happened in Utah as the Mormon war and the death of the prophet.[4]

The Morrisites, like the Mormons, were preoccupied with the second coming of Christ. They believed that the discovery of America was a sign from God that the time had come to renew the Christian world, and that America had been chosen above all other lands as the place for the second coming of Christ and the restoration of paradise on earth. Joseph Smith had taught that the establishment of the City of God would be in the United States. The Morrisites believed that the second advent of Christ would take place in their midst. They expected to play a leading part in that society and worked diligently to prepare themselves to be worthy participants. On August 9, 1865, Cainan wrote to John D. Eardly, "As Deer Lodge Valley, Montana is the nucleus, let all others begin to turn their attention that way..." This was Cainan's call for the migration which took place in 1871. He wrote later to William James, (Jan 6, 1879) "For the choice place of the earth is handed over to you...For there will be built the City of the Great King."[5]

One can easily imagine the band of Morrisites weary and foot-sore, aggrieved refugees from slaughter, bloodshed and hopelessness, but firm in their faith of the Second Coming, stopping in wonder as they gazed on the Deer Lodge Valley spreading out before them. Wasn't this the promised land, wasn't this where God's Kingdom on earth would be established? All they saw could only affirm Cainan's prophesy. In the distance, slate blue mountain tops with snow-white caps trailed each other across the horizon like giant biblical elephants partially hidden in the tall grass of ever-green foothills. They marvelled at the beauty of it as the evening sun dropped behind the tallest peaks transforming the landscape from three dimensions to two, flattening the mountains so that they became black jagged silhouettes pasted on a dark blue sky turning crimson with occasional puffs of pink, white and gray clouds. Wasn't the stillness of the moment, accented by the clear, dry, rarified mountain air, charged with just a hint of eternity?

Couldn't these bastions of the eons, guardians of permanence and continuity offer these refugees from chaos and confusion not only the magic of that moment, but a lifelong promise of certainty, peace and harmony? Surely it was God's will that they settle in the folds of this magnificent range.

Anyone who has contemplated this scene in late afternoon and watched as the shadows spread from the valleys across the wooded foothills, darkening and flattening their contours transforming them into a gigantic dark wall against the sky cannot be but awestruck by its silent grandeur. If, indeed, God is to establish His kingdom on earth, why not here?

But what was the attraction of this valley for its other early settlers? Besides its natural beauty, it was blessed with abundant water, always a priority in the arid West. The valley is a large depression, that stretches for sixty miles between two mountain ranges, and is surrounded by wooded hills and snow capped mountains. At its widest, the region that the Morrisites saw when they came over the hump from Butte, it is eight to ten miles across. Off the main valley are many smaller valleys and canyons each with fast-racing, cold, clear mountain streams, fed during the dry summer months by the melting snows of the high mountains. There are also many natural springs, some of them hot. The land is fertile even though the growing season is short due to its northerly latitude, and to the high altitude, over 5,000 feet above sea level.

Before the valley was settled one of its most striking features was a natural formation, a huge conical mound, over forty feet high, with a thermal fountain bubbling warm water out the top and down its sides. When the surrounding air was cooler than the water, which was most of the time, a clearly visible cloud of steam issued from its summit. From a distance it looked like a huge Indian lodge with the smoke of the camp fire rising from it. This formation gave the valley its name. The Indians called it the "Lodge of the White Tail Deer," for the abundance of white tail deer that fed on the almost always green grasses in its vicinity. By extension, the valley was named after its most prominent feature. That spot today is called Warm Springs, and the clear, cold mountain stream that flowed by that spring was named Warm Springs Creek. On the valley floor Warm Springs creek flows into the Deer Lodge river which is formed as Silver Bow Creek wends its way north and is joined by the waters of Willow Creek, Mill Creek, Warm Springs and Lost Creeks.[6]

Warren Ferris, a fur trapper, said in his journal that he had come to "the Deer House Plains" and found 100 lodges of Pend d'Oreilles camped in the area with 3,000 horses. The largest gathering recorded occurred August 18, 1854, when 6,000 Shoshone Indians arrived for recreation and to settle grievances among the many bands.[7]

It appears that the warm waters flowing from the mound's summit had already ceased by 1840 when Fr. DeSmet, stopped in the valley. On his way to minister to the Flathead Indians in the Bitter Root he wrote: "By far the most remarkable spring we have seen in the mountains is called the Deers' lodge...The water bubbles up and escapes through a number of openings at its base...and ... are of different temperatures: hot, cold, and lukewarm though but a few steps distant from each other. Some are so hot that meat may be boiled in them We actually tried this experiment." [8]

Information on precise dates when the Indian name, in its English translation, came into general use is sketchy. It could have been soon after 1810 when representatives of the British-owned Hudson Bay Company and John Jacob Astor's American Fur Company began rendezvousing in the valley with the Indians and French and American trappers to buy furs. By 1831, the name began to appear in trappers' journals. A Warren Ferris map of 1836 shows the mound, and the DeSmet visit of 1840 is recorded. Most of the valleys in the area had already been named in 1853, when Isaac I. Stevens, commissioned by the US government to survey a railroad route from Lake Superior to Puget Sound, left a survey party, under Lt. John Mullan, to explore and map the Bitter Root, the Big Hole, Deer Lodge and Beaverhead valleys. Stevens called the mound "a remarkable curiosity." Mullan, plunged a 20 foot pole down the throat of the mound "without reaching bottom." The Deer Lodge Valley and the Deer Lodge River were well established names when the Montana Territory was formed in 1864 and the county came into existence with the same name. The town of Cotton Wood on the Deer Lodge river became Deer Lodge in 1861.[9]

About 1871 Louis Belanger, a French Canadian from Quebec who had followed gold from California to Montana, sold out the general store he was running in Silver Bow and homesteaded a ranch site at Warm Springs, which included the springs and the mound, where he began raising cattle. In subsequent years he built a two-story hotel of ten rooms and several bath houses. In 1875, title passed to Dr. Charles F. Mussigbord and Dr. Armistead Mitchell. They promoted the natural beauty of the area as well as the "health giving and renewing waters" of the springs to secure a contract with the Territorial government to care for the mentally ill. The institution opened with thirteen patients for which the territorial government paid them $1.00 a day per patient.   The partners gradually expanded their holdings so that by 1886 they owned approximately 6,800 acres.[10]

Settlement west of the future Anaconda townsite began when Peter Levengood started ranching there in 1865. He had left Missouri the year before bound for the gold fields at Virginia City. After working the placers there for about a year, he moved to upper Warm Springs Creek and sent for his wife and children. After a trip of eleven weeks up the Missouri river from St. Louis to Fort Benton, then overland to the Deer Lodge Valley, they arrived in 1867. The Levengoods were the first white settlers to remain in the upper Warm Springs Creek area. A few years later Jake Stuckey, another ex-prospector, became Levengood's neighbor, homesteading 160 acres nearby. Levengood established the Levengood post office and was well known by the miners at Cable, Southern Cross and Georgetown. In 1883, he sold part of his ranch to Marcus Daly for the Anaconda townsite.[11]

Another gold seeker turned rancher was, John P. Thomas, son of William P. Thomas, who had accompanied his parents to Utah in 1850, as a boy of seven. He arrived at Bannack in 1863 hoping to strike it rich. In 1865 he came to Deer Lodge Valley and bought a ranch just west of his father's on upper Warm Springs creek where he raised cattle for two years, before moving on to Idaho, for another try at mining. He returned to the Deer Lodge Valley in 1871 to raise sheep, before moving again to Butte where he opened a grain store. After farming for a number of years in the Judith Basin, he returned to the then growing town of Anaconda, built a house and became a permanent resident.[12]

Another early arrival in the area was Charles Bingley Jones, from Indiana, who drove to Corrine, Utah by team and arrived in Virginia City in 1863. He tried hauling freight between Salt Lake City, Virginia City and Fort Benton, then logged for two years in Madison County before starting a ranch on upper Warm Springs Creek in 1867, which, sixteen years later, gave way to the rapidly expanding new town of Anaconda. In 1887, Mr. Jones married Minnie Hartley, who had come west with her parents by ox team to settle in the Deer Lodge Valley in 1865.   Both Jones and Jake Hartley, Minnie's father, who had bought the John Thomas place, sold out to Marcus Daly for the new townsite.[13]

In 1871, Daniel Murphy, an Irishman, settled on land which today is part of the village of Opportunity, five miles east of Anaconda. He had arrived in Alder Gulch in 1863, tried his hand at placer mining, then moved on to Silver Bow Creek and later French Gulch working the placers before settling in Deer Lodge Valley. In 1880, he homesteaded 160 acres where he had settled and later bought an additional 40 acres from the Union Pacific Railroad.[14]

John D. Harris, born in Syracuse, Ohio, traveled with his Welsh parents, Isaac and Mary Harris, in a covered wagon from Ohio to Utah, then to Virginia City before settling in the Deer Lodge Valley in 1873, near the stage stop at Stuart. They joined a small colony of other Welsh immigrants who had settled there.[15]

William Beal, a stone mason, with his wife and two sons left Fayetteville County, Pennsylvania, in 1874. They traveled by railroad to Corrine, Utah and then by covered wagon to Bannack. After helping build a hotel there, he and his family moved to the Deer Lodge Valley and homesteaded on Warm Springs Creek on land now part of the city of Anaconda. He subsequently moved further downstream to land which later became part of the state mental hospital at Warm Springs.[16]

T.C. Davidson left Missouri by ox team and arrived in Butte in 1879. After hiring out to haul silver ore from the mines for about a year, he left Butte and bought a ranch on Warm Springs Creek, raising cattle and farming. In 1883 he sold his ranch to the new Anaconda Company as part of the site for its smelter. Davidson later built one of the first commercial blocks in the new town of Anaconda.[17]

While most of the newcomers to the area settled in the valley to farm and ranch, others more interested in finding gold followed Warm Springs creek west into the mountains. One of these was Frank Brown, an old prospector who discovered silver in 1875 about five miles west of where Daly was to build his new smelter eight years later. Brown called his mine the Blue Eyed Nellie after his young daughter, who died shortly after he had hit pay dirt. With financing from W.A. Clark, he sank a five hundred foot shaft, built a tramway down the mountain from the mine and a fifty-ton blast furnace on Warm Springs Creek to smelt the ore. Later, when Daly's smelter was operating, the new railroad was extended into Brown's Gulch to haul out Brown's ore and also lime rock as flux material for the smelter. This site later became known as the Lime Quarry.[18]

While the majority of settlers in the valley came from the East via Colorado, Nevada and Utah, some came from the West Coast attracted by the gold finds in Idaho. One of these was Henry Harrison Eccleston from Junction City, Oregon who found his way to Stuart, where he arrived in the early 1870's. After making a small stake at mining in French Gulch, he moved to White Pine Creek just south of Gregson where he started a ranch and raised his family. Eccleston, like many others at the time, mined and ranched intermittently.[19]

Of course, not every settler in the valley arrived via the gold camps. The Staton family had come to Utah from Missouri with a Mormon wagon train in 1871. In 1882 they decided to move on to Oregon. They went through Idaho and into southwestern Montana. As they followed Silver Bow Creek west, past the abandoned placer diggings and the ghost town of Silver Bow City, they came over the hump from Butte, and saw a marvelous expanse of green valley spread out before them. The tall grass rippling in the wind resembled a vast ocean. The afternoon sun reflected off the numerous streams which cascaded down the heavily wooded mountains surrounding the valley then wound across the flat and joined to form the Deer Lodge River. Peter Staton decided this was a perfect place to rest for a couple of days. They stopped at a homesteader's cabin about seven miles east of the present town of Anaconda. After some discussion, the homesteader agreed to trade his cabin and 160 acres of land to Peter Staton for a team of horses and a wagon. The Statons stayed and raised a family on that farm. They had 50 acres of vegetable garden and 80 acres of hay. The miners in Butte made a good market for the Staton vegetables.[20]

Unlike the white man, the Indians were attracted to the valley not to search for gold, nor to seek eternity. Rather it was those white tail deer they prized - - as well as beaver, elk, moose, the fish in the streams and the health-giving warm and hot water springs. They made no permanent settlements here, but considered the valley a neutral area to be used by wandering bands from many tribes. It was frequented by the Blackfeet, Crows, Pawnee, Shoshone and Snakes and was a favored camping spot for the Flathead, Pend d'Oreilles and Nez Perces tribes on their annual treks eastward "on their way to

Buffalo" and in search of food and game. Like other nomadic cultures that live off game, berries, nuts and roots, while the braves moved east into buffalo country, they would leave some of their women there in eight to ten lodges to forage until the braves returned from the hunt.[21]

This routine did not change with the arrival of white pioneers, who settled in groups for protection but became accustomed to the annual Indian migrations through the valley. Old Chief Joseph and his Nez Perces tribe were known to the new settlers in the valley before the famous Battle of the Big Hole in 1877, and although Deer Lodge Valley provided a contingent of volunteers for that battle at the urging of the Territorial Governor, the settlers were much less apprehensive about this tribe than the U.S. military bent on forcing them to stay within their reservation in Idaho.[22]

There is only one vague reference to possible conflict between whites and Indians in the valley and it is related to a hill at the mouth of Lost Creek on the north side. It is a well rounded hill coming to an abrupt point at the top and referred to by old timers in the valley as "Squaw Tit Mountain." For years visitors could see the remains of a wall or ring of boulders on the top of it, which seemed to indicate a grave or graves. Some long-time residents remember stories about a battle between the Nez Perces and the Shawnee in which three Indian chiefs were killed and buried on the mountain top.

Another version, related at some length and with great rhetorical flourish in the *Montana Standard* in 1941, told the story of a group of twenty-four prospective gold miners, with seventy-five horses, who were making their way from Canada to Virginia City in 1863. On Oct 13, they were attacked by a war party of over one thousand Crow Indians. The Crow had been pursuing a large band of Gros Ventre Indians with whom they were at war who had passed that way earlier in the day. Attracted by the gold prospectors' string of horses, they temporarily abandoned their chase of the Gros Ventre and attacked the whites instead.

Led by one Donald MacKenzie, the gold mining party had camped on Lost Creek where it cuts through the foothills into the open meadows of the Deer Lodge Valley. They spotted the Indians across the valley early in the morning as they emerged from Silver Bow Canyon. Armed with breach-loading Winchesters and prepared to defend themselves, the prospectors climbed squaw tit mountain and

built a stone fortification. After trying to take the hill all that afternoon and into the next morning, the Crow finally collected their dead
and withdrew. The MacKenzie party suffered four killed and eleven
wounded during this encounter. The dead were buried where they had
fallen, on top of the hill. Twelve years later, Chris Johnson, who had
been scout for the group returned to the spot and placed headboards
over the graves, one each for: John Marks, English; Reil Renquist,
French-Canadian; Hans Swieger, German; and John Forbers from
Ohio.[23]
    As rich as this story is in detail, with names of Indian tribes
and individuals as well as dates, it is strange that it escaped the
attention of the original settlers in the valley, none of whom ever
mentioned such a battle and some of whom clearly believed that the
ring of stones on the hilltop marked Indian graves.

    By the time Anaconda was founded most tribes were con-
fined to their reservations and the annual trek through the valley was
only a memory. But in 1885, after the Riel rebellion in Canada, Anaconda and Butte attracted a tribe of outcast Chippewas, refugees from
that conflict. The Chippewas, originally from Wisconsin, had come
into Montana from Canada hunting buffalo along the Milk River.
Under their chief Rocky Boy, they set up camp in Deer Lodge Valley
and frequented the new town of Anaconda, "begging, stealing, selling
cayuses, hauling wood, and making an odd dollar selling horn hat
racks, chairs and ornaments, one of which could be found in every
parlor in the early days."[24]

    A 1902 newspaper story gives a flavor of some of the
changes that had taken place in Indian life in the valley:

"Feb. 5, Bannocks, Crees, Snakes and Flatheads gathered a few
miles east of Anaconda last night, where the Chippewa camp
was putting on a dance. There were about 200 of the Chippewa
present and participating. Decorated with feathers and paint,
they executed the old-time Indian two-step so popular with the
red men when they were the only bosses of Montana. When it
became too cold out of doors, they set up two huge tents. As
an open fire in the dancing pavilion was impractical, because
of smothering smoke, the Indians acquired a stove from their
white neighbors. The dance music was supplied by six Indians
who stretched a piece of buckskin over a zinc washtub and
then pounded the skin with small clubs. A reporter present

turned to a lavishly-decorated Bannock who sat watching with stony expression of countenance, the evolution of the dancing bucks and squaws, and said: "You speakee English, chief?" "Oh, in a slight measure," he replied. "I came up from Pocatello to witness this dance and I feel that I have been amply rewarded. If you desire to quote a Bannock on a Chippewa ceremony, you may say that introduction of a heating stove into an Indian dancing pavilion is a decided innovation. Next year they will undoubtedly have steam coils set up along the walls."[25]

Even this reminder of the valley's Indian past disappeared in 1916 when the Chippewas departed for a new reservation in Northern Montana, which had been created for them and the Crees, also wandering refugees from Canada.

3. Boom Town

Before Daly's momentous discovery, many of the silver mines on the Butte hill had run into copper. Most considered it a nuisance and paid little attention to extracting the red metal from the ore. Even so, by 1880, there were already three primitive copper concentrators and smelters operating in Butte. The Parrot Silver and Copper Company, owned by Andrew J. Davis and Samuel Hauser; the Colorado and Montana Company, owned by Nathaniel Hill and W.A. Clark; and the Montana Copper Company, owned by Adolph and Leonard Lewisohn, all produced both silver and copper ores. Most ores, however, were sent for processing to smelters in Colorado and Baltimore. Some was even shipped to as far as Swansea, Wales, then a world metallurgical center. Until 1881, when the Utah and Northern built a narrow-gauge railroad spur to Butte, this meant trundling ore by oxcart from Butte to the railhead in Corrine, Utah, and thence by rail to an east coast port for shipment across the Atlantic or to Portland Oregon and thence by sailing ship around the horn of South American to Wales.[1]

The price of copper was a healthy 18 cents a pound in 1881, world copper consumption at the time did not warrant large-scale production. Moreover, the ore mined in Butte would have to compete with the native, practically pure, copper from the Lake Superior copper range on Michigan's upper peninsula. Daly and his partners in the syndicate were probably only vaguely aware that their discovery in Butte coincided with Thomas Edison's invention of the electric light bulb and the inauguration of the first plant for the generation and transmission of electric power in the United States in New York City, which set in motion an exponential expansion in the demand for copper wire. Nevertheless, they went forward mining the new vein and shipping the ore as far as Wales for smelting.

When Daly's first shipments assayed out at 55% copper, the people in Swansea couldn't believe that they were receiving raw copper ore and wanted to know what concentration process had been used prior to shipment. They had never before seen or heard of the kind of massive ores now coming out of Butte. Of course, because of the high transportation costs, Daly was shipping only his highest grade ores. The average copper content of ores from the Anaconda mine ran at a more modest 12%. Nevertheless, from 1882 to 1884, Daly shipped 37,000 tons of ore averaging 45% copper content from Butte to Swansea.[2]

In 1883, the syndicate's gross earnings had increased to $1,702,400 from copper and $1.6 million from silver, through the Swansea shipments. Satisfied now that the ore from the Anaconda mine could be successfully processed, Daly convinced his partners that the time had come to build a smelter of their own. He had already been buying up land at his intended site for over a year.[3]

The Haggin, Hearst, Tevis mining syndicate had already racked up spectacular successes in gold and silver. They owned the world's richest gold mine, the Homestake in South Dakota, and the world's richest silver mine, the Ontario in Utah. Now they were prepared to bet, with substantial new investment, that their recent acquisition, the Anaconda in Butte, would become the world's richest copper mine.[4]

James Ben Ali Haggin, the leader of the syndicate, believed in the long shot and was not afraid to spend whatever was required to reach his goal. He had full confidence that his partners in the Anaconda enterprise, Hearst and Daly, had correctly appraised the potential of their Butte property so gave Daly a free hand to build a new smelter.

From the moment he picked up that piece of copper glance, Daly realized that he'd eventually have to build a smelter, so he began scouting around for a site. The Butte *Daily Miner* newspaper speculated on June 24, 1882 that the Anaconda Copper and Silver Mining Company would erect a smelter in Butte. But Butte would not do. It lacked sufficient water and the three small smelters already there spewed out smoke that covered the town in a permanent haze.

After examining locations on the Madison and Big Hole rivers, Daly decided to build his new smelter on the hills just north of Warm Springs Creek in upper Deer Lodge valley twenty six miles west of Butte. The town to support the smelter would be built on the valley floor on the south side of the creek.[5]

The extraction of metals from hard rock was still in its formative beginnings in the United States of 1883. Gold fever, which had driven tens of thousands of fortune-hunting prospectors to risk life and limb, suffering untold hardships to explore the remotest streams and riverbeds across the West, also initiated a technological revolution in mining and the extraction of metals from the mined ore.

The early prospectors were placer miners, individual entrepreneurs washing sands from streams and river banks in pans, sluice boxes, rockers and other simple devices which washed off the lighter elements and left the heavier gold in the bottom to be gathered up with mercury which was later evaporated out. Getting the gold out of gold-bearing rock mined from veins running under ground required greater effort, organization and technical skill. To break and grind the rock so that the ore could be further worked the earliest operators utilized the old Spanish "arrastra". This consisted of a hard surfaced circular bed around which large stones attached to a central pivot were dragged by mules, oxen or a water wheel. The pulverized rock was wetted down to form a slurry, then copper sulphate, salt, and mercury were added to amalgamate with the gold. A variation on the "arrastra" was the Chili wheel which use a heavy stone wheel in place of the large stones. Another method was the stamp mill which pounded the rock into powder in a huge mortar. The stamps were heavy iron heads attached to shafts and powered by steam. Smelting was not required for gold or the Nevada silver ores. But copper was a different story. Since the Civil War, the Lake Superior copper producers had been using a concentration process for some of their lower grade copper ores which eliminated the waste rock and left only the valuable residue to be transported to a smelter. The concentrators were ball or roller mills which pulverized the ore, mixed it with water and passed it through agitation and settling processes.[6]

The shift from placer to quartz mining meant that the lone prospector with his mule and grub-stake was already giving way to an industrial operation that required large, long-term capital investment, technical knowledge and hired labor. Butte had seen this transformation from the gold placer operations along Silver Bow creek to the mine shafts on Butte hill in search of silver. Daly, himself, had witnessed it in Nevada and later in Utah. He had learned that a successful mining operation required massive capital investment and the most up-to-date methods and machinery. Since the copper sulphide ores required roasting, concentration and smelting, Butte, even at the time of Daly's arrival, was transforming itself from the typical mining camp to a new phenomenon in the West. It was becoming an industrial center on the edge of the frontier with smoke-belching open-air furnaces roasting the sulfur out of the silver and copper-bearing sulphide ores and primitive mills and smelters concentrating and melting down the residue. The hills around the mines were denuded of grass and trees and the camp itself lived under a continual haze of sulfur and arsenic smoke. Daly meant to roast, mill and smelt his ores in a separate location.

He commissioned his friend Morgan Evans to begin quietly buying up land for both the smelter and the town. By November 1882 Evans had filed claims for water on Warms Springs creek and Mill Creek for industrial purposes. He first bought land owned by Gordon Vineyard northeast of the townsite on Warm Springs Creek. He then bought the Nate Levengood ranch and secured the rights to Northern Pacific railroad land from Charles B Jones and Jacob Hartley where the town was to be built. He also bought land west and north of the townsite from Robert Finley and Alexander Glover. By May 1883, 3000 acres of land had been acquired. The total cost for these lands came to $20,000, but this was only the beginning of Daly's purchases along the creek. By June 1896, a Company report indicated that his land and water right purchases totaled $401,475. [7]

Even before the city plat was filed in Deer Lodge, on June 25, 1883, work had begun in May on the excavation of a large ditch to carry water from Warm Springs Creek to the new smelter, later to be called the Upper Works when another smelter was built further downstream. Tents stretching along the creek already sheltered the growing work force, which began making its way from Butte and other points near and far.

At about this time, filings were made by Marcus Daly, William McCaskell, Haggin's expert works superintendent from San Francisco, and William Read, formerly with the Walker brothers and now Daly's accountant, with the Clerk and Recorder of Deer Lodge county, for the location of mill sites, placer claims, rock quarries and water rights. Morgan Evans filed for 500 inches of water from Warm Springs Creek "at a point near Alex Glover's house for the purpose of milling, concentrating, smeltering and manufacturing, and to be used as motive power and washing ores." The water was to be taken to the works, used and returned to the creek "opposite the mouth of Morgan Evans agricultural ditch starting from a point on Vineyard's old ranch." Evans also filed for 5,000 cubic inches of water to come from Warm Springs Creek opposite Jake Stuckey's farm (2 to 3 miles west of the town) and 2,000 cubic inches from Mill Creek between Allen's and Parker's ranches. All of this for industrial purposes.[8]

Rock and large stones for the foundations of the new smelter were quarried from the foot hills on the south side of the valley from the "Sheep Canyon Placer Claim", later referred to simply as Sheep Gulch, named for the herds of big horn sheep that frequented the area. As the work progressed, a narrow-gauge railroad spur was built across the upper townsite and up Sheep Gulch to the quarry.[9]

Daly hired James Harper to survey the townsite and contracted with Daniel Dwyer, known for his outstanding masonry work in Butte at the Anaconda mine, to work with stonecutter Thomas Burke and crew to build foundations for the new smelter. The Vineyard ranch became company headquarters and the old T.C. Davidson house, bought by Daly in 1882, became the company time office.[10]

Most of the materials to build the town and the smelter had to be hauled in. Hundreds of freight teams hauled timber from the hills north of the smelter site, from Lost Creek, Mill Creek, and Willow Creek and still others brought in supplies from Butte. A wagon road was built from the Utah and Northern Railroad station at Stuart across the valley and up Warm Springs Creek to the smelter and town site. A stage company was established to meet the trains and transport passengers from Stuart to the new town.[11]

The mines in Butte had a voracious appetite for new timber in the form of stulls and props to shore up the shafts as the silver and copper veins were followed deeper into the earth. The hills around Butte had been stripped clean even before Daly arrived there. Mineowners had to go further and further afield to satisfy their timber needs, estimated at between 40 and 50 million board feet a year. Railroad building, with its increasing need for wooden ties, and wood burning steam engines added to the general need for timber products. Building the smelter in Anaconda compounded this demand and, as in Butte, the forests nearest to the smelter site were the first to go.

In 1882, before work on the smelter began, Daly sought wood as far away as Missoula, where the Northern Pacific line was being built. He formed the Montana Improvement Company, in partnership with the Northern Pacific and lumbermen A. B. Hammond, R. A. Eddy, and E.L. Bonner, primarily to ensure that he had a steady supply of timber for his mines. This company also engaged in constructing railroads, canals, docks and waterworks, and had a contract to supply timber and ties to the Northern Pacific Railroad.[12]

Rather than use his Missoula Company, Daly advertised for a separate contract to take care of the fuel needs for his new smelter and lumber for the town. The huge amounts required in his request for bids stunned local observers. He wanted 300,000 cords of wood at the rate of 75,000 cords a year. Someone calculated that if that many logs were placed in piles four feet high, they would reach from Butte all the way to Salt Lake City. Accepting the lowest bid of $3.49 a cord, he signed a contract with John Caplice, Albert W. McCune and Captain John Branagan. The timber was to be brought out of Mill Creek canyon southeast of the town. It was a massive operation utilizing eighty teams of horses and mules and requiring 650 men. Logs were cut and snaked out of the forest by mule teams then transported by water down wooden flumes, constructed for the purpose, to the mouth of Mill Creek canyon, where they were brought by wagons to the construction sites. When the railroad was built to Anaconda, the flume was extended to the rail line.[13]

Later William N. Allen, who had a ranch on Mill Creek, took over this contract under the name of the Allen Timber Company and continued providing timber primarily for the Butte mines. As the nearby forests were depleted, cutting moved further away and the flume was eventually extended fifteen miles over the continental divide into French Gulch on the way to the Big Hole. Thus waters were diverted from the Atlantic slope through the flumes to the western side of the divide and would eventually wend their way to the Pacific ocean. With timber cutting, the dwindling gold camp population of French Gulch increased dramatically. Allen had between 500 and 600 men on his payroll cutting timber for stulls for the mines.[14]

The work force for the smelter was recruited first in Butte, where miners and prospectors, attracted by the lure of silver, were still arriving from Nevada, Utah and Colorado. But Daly cast a much wider net, recruiting from eastern Canada and the copper towns of Houghton and Hancock on Michigan's upper peninsula and as far away as Wales. He needed specialized skills and was willing to pay whatever it took to attract them. He brought in Hugh MacMillan and a group of Scotch millwrights from New Brunswick and Nova Scotia,

Canada, to frame the timbers on the first units of the Upper Works. Bob Vine in his excellent "Anaconda Memories 1883-1983" says there were 62 Macs on the construction payroll and so many Hugh MacMillans that they had to be listed with prefixes such as Big Hugh, Little Hugh, and Red Hugh to distinguish them. Daly also hired French Canadian lumbermen who were good with a broadaxe to work in the woods. News about the construction of a new smelter in Montana also reached recent arrivals from Europe looking to start a new life in the United States.[15]

Thus, Anaconda sprung from many sources. Early settlers included American-born pioneers from places further east and Mormon dissidents of English, Welsh and Scandinavian extraction. There were also Irish immigrants, who like Daly had made their way from mining camp to mining camp; Welsh recruited for technical jobs in the new smelter; Scotch millwrights from Canada; French-Canadian lumberjacks and recent Scandinavian immigrants, Swedes and Norwegians, who arrived by way of Michigan, Wisconsin and Minnesota, and even a handful of Chinese. They were soon joined by a small number of Germans who manufactured and sold cigars and beer, or went into commerce and a scattering of Italians, who came to work on the smelter.

Matt Kelly, in his "Anaconda, Montana's Copper City", a wonderful rambling tale of the early years, says the stage company running between Stuart and the new town reported in March 1884 that from 20 to 40 passengers arrived each morning by rail and took the stage. He includes the following quote: "We now have two lines of stages so that parties coming on any of the day trains have no trouble in securing transportation to the future metropolis of the Deer Lodge Valley. During March, we hauled from Stuart Station, over 460,000 pounds of merchandise. Kirkendall and Brown run the freight transfer line which carries daily about 20,000 pounds".[16]

The first inhabitants of the town, lived in tents along Warm Springs Creek. They were a motley mixture of rough and tough, hard-working, hard-drinking men, without their women, and were used to the wild and makeshift life of the mining camps. So from the beginning, the new town took on the air of a mining camp. Bob Vine quotes an account by David Marler recalling the town's first Fourth of July celebration:

"James Keefe had three tents, which were doing duty as a boarding house. Further west were two more tents, both of them used as saloons.

"Having nothing else to do but drink, nearly the entire population began to tank-up early and kept at it all day. Fearing a wholesale shooting scrape, the few men who kept sober went around collecting the guns of those who were drunk, and it is well that this precaution was taken, or there might have been a dozen bloody murders."[17]

Even though the emergent town showed signs of becoming a smaller version of bustling, rambunctious Butte hill, Daly had other ideas for his new queen of cities. The town would be separate from the smelter, not like Butte where miners shacks and commercial establishments grew up around mine headframes and mine yards. It would be a planned community laid out in a classical grid pattern, on the south side of the valley far enough from Warm Springs Creek for a railroad depot and the accompanying railroad yard, but still close enough that workmen could walk to and from their jobs. This would be a neat, orderly town, of family homes with small gardens and tree-lined streets and a clearly defined business and commercial district. An early city directory describes it this way: "From end to end Anaconda wears a tidy air. Its streets are clean, its homes are cozy, and its dooryards are well kept." It would be a place that Daly could build a big house fit for his wife and four children to live.[18]

Daly's mine was named Anaconda and so was his company. What about his town? It is reported, and has become legend, that he wanted to name it Copperopolis, but Clinton H. Moore, the town's first postmaster, in filling out the application for the new post office, wrote "Anaconda" into the space for its name.

Moore, a native of New Hampshire and a graduate of Dartmouth college in 1874, came west and was hired as principal of the new College of Montana at Deer Lodge in 1878 and was Superintendent of Schools in Boise Idaho before being named postmaster of the new post office in 1883. With regard to the naming of the town, Moore has been quoted as saying that he told Daly there was already a camp and a post office named Copperopolis in eastern Montana and the Post Office Department did not like repeating names. Even though Daly continued to favor Copperopolis, Moore named the post office Anaconda.[19]

Matt Kelly insists however, there was never any real dispute about the name since Morgan Evans filed the city plat June 25, 1883 with the name "Anaconda." Also, all the filings for mill sites, placer claims and water rights were filed in May and June with the name "Anaconda". Kelly says it was four months later, October, 1883, that Moore wrote "Anaconda" in the blank on the post office application form. County records indicate the new township of "Anaconda" was created June 26, 1883, confirming Kelly's contention.[20]

Anaconda, or Copperopolis, whatever her name was in the summer of 1883, became a thriving, exciting, vibrant boom town almost overnight. City plans were drawn up and streets laid out and named. Frame buildings for businesses and dwellings sprang up like mushrooms after a rain. J. B. Keefe opened the Alamo Hotel on the corner of Main and Front streets. David Cohen built a store on Main street between Front and Commercial. Jim Hamilton had a grocery store next to Keefe's. J. Ross Clark, Daly's brother-in-law, built a store on the corner of Commercial and Main. William Read opened the first drug store at 17 Main Street. William Jack of Butte opened a store managed by Harvey Mahan. Chris Kessler built a frame building and Bob Fenner opened a brewery on Third Street. Patrick Maloney, who had learned tailoring in New York City on his arrival from Ireland and had been conducting his business out of a tent, opened a tailor shop and sent for his wife and five small children to join him. Two years later Maloney would form the town's first volunteer fire department. Jim Douthat built a boarding house on Front street and Jack Morris, who had come from Butte with Dan Dwyer under contract with Daly to lay the foundations for the smelter, built another boarding house further east on Front, near Cedar where the wagon road crossed the creek to the smelter.[21]

In August 1883, about two months after the first tents were pitched, a business survey showed that Anaconda had three brick-yards, several sawmills, a sash factory, a lumberyard, blacksmith shops, boot and shoe shops and grocery stores. The population was about 1500. The town was also adding to its ethnic diversity. In 1883, Sam Pramenko, a Serbian from Herzgovina arrived in Anaconda to open a grocery store with a saloon in the back. Sam had come from the silver camps of Nevada, where he had also run a saloon and grocery store. For a short time he ran the store with his countryman, Mike Ivankovich. A couple years later George Barich, from Croatia, who came to Anaconda via Pennsylvania and Denver, also opened a saloon and boarding house then started a clothing and feed store. Both of these stores were on First Street, later called Commercial Street. Barich subsequently built one of the first blocks in Anaconda on Park Street, where he operated his feed store and saloon. These men provided encouragement and a base for others of their country-men who would arrive in subsequent years. The job opportunities in Butte and Anaconda were such that by the end of 1883, the popula-tion of the territory had increased from 40,000 in 1880 to 143,000.[22]

Not only laborers, tradesmen and merchants were attracted to the area, but professional men as well. In October, 1883, Dr. Oliver Leiser arrived in Anaconda and opened an office on the corner of Commercial and Oak streets, becoming, along with Doctors Snyder and Mitchel, one of the town's first physicians. A native of White Deer township, Pennsylvania and a graduate of the University of Pennsylvania, Dr. Leiser practiced first in Helena with his brother, then came by horseback and dead-axle wagon to Anaconda. He later became head doctor at the town's new hospital and in 1897 became the town's mayor. He also served as county coroner and was a member of the state legislature.[23]

In early 1884, George B. Winston arrived to open a law office. A native of Missouri, he came to Butte in 1883 and moved to Anaconda soon after he was admitted to the Montana bar. He later became Anaconda's first City Attorney, first City Clerk and was one of the seventy-five men to assemble in Helena in 1889, to write a constitution for the new state to be. Winston later became a judge and established a record for longevity on the bench, serving from 1904 to 1933. In 1896, he hired Bridget Sullivan, a maid who had worked for the notorious Lizzie Borden. Bridget worked for Winston until he died in 1936, never once mentioning the gory details of what she had seen that hot summer day in 1892 in Fall River, Massachusetts when her employers, Andrew Borden and his wife, were killed with an ax allegedly wielded by their daughter, Lizzie. Bridget married John M. Sullivan in Anaconda in 1905 and lived with him at 701 Alder Street for thirty-four years. When John died in 1939, she moved to Butte where she lived until her death in 1948. Both she and her husband are buried in Anaconda's Mount Carmel Cemetery.[24]

The town's first bank, Hoge, Daly and Company, with Marcus Daly as president, W. L. Hoge vice president and William

Thornton as cashier, opened in September, 1883 in a one story brick building on the corner of Commercial and Main streets. This was one of the first brick buildings in the town. The bank immediately began to sell city lots for Daly's Township Company. A corner lot on Main Street sold for $750 and $500 for the others. Those on Front Street went for $350. Hoge also built a double row of houses, known as Hoge's row, on east Commercial Street. One of them he donated to become the town's first school which opened in October. Apple boxes were used to sit on and write on until proper desks could be made. Before that the few children in the town had attended a small country school near the Levengood farmhouse. There was another country school about five miles east of the city near Gwendale, the Evans ranch.[25]

By late 1884, the town had eighty houses, including seven hotels and boarding houses, twelve saloons, six stores and one butcher shop. There was a telephone exchange switchboard installed in Harvey Mahan's store which had ten subscribers. That number increased to 18 the following year. There were over 1200 men on the Anaconda Company payroll, most of them single, living in boarding houses. By 1885, the number of men on the payroll had increased to 1700. It is interesting to note that, at this time, Daly had more men on his Anaconda payroll than he did in his mines in Butte. His Butte payroll was running at $100,000 a month for 1500 men, while the monthly payroll in Anaconda was $140,000. [26]

While Daly was constructing his smelter, the Reverend E. J. Stanley and his congregation of the Methodist Episcopal Church South began building the town's first church. It was completed in 1884, and stood on west Park Street, across the alley from where the Montana Hotel would later be built. It was a brick church with a steeple topped by a weather vane in the form of a fish. Shortly thereafter the Christian congregation built a frame church on the corner of Cherry and Park streets. These two buildings, referred to as the brick church and the frame church, were also used by other congregations until they could erect buildings of their own.[27]

The town's first newspaper, *The Anaconda Review*, was started on May 1, 1884, by John S. Mills. It was printed in Butte until a new owner, L. O. Leonard and his brother Frank, bought a used hand press and ran off the first newspaper printed in Anaconda on July 23, 1885.[28]

4. One Hundred Years of Ore

Daly needed a railroad to transport the ore from Butte to his new smelter and to send his smelted and refined copper to market. So, shortly after he started construction of the smelter he began negotiations with Jay Gould and Stanley Dillon, of the Union Pacific Railroad, whose narrow-gauge Utah and Northern spur ran from Butte to Garrison. Dillon was skeptical believing that Daly's Anaconda would become just another ghost town as soon as the ore vein in Butte ran out or the price of copper fell. Daly, however, in his usual optimistic and expansive way, convinced Dillon's representative, Mr. E. H. Wilson, by telling him that there was enough high-grade ore in his Butte mines to keep the Anaconda smelter and the proposed railroad busy for the next one hundred years.[1]

Only future generations would come to appreciate how prescient that prediction was. Ore would roll over railroad tracks from Butte to Anaconda for the next ninety-seven years. But at the time Daly was trying to make a point not a prediction. He needed an eight and one half-mile track from Anaconda to the Utah and Northern line at Stuart. By April 1884, work was well advanced and in July the job was completed.

The *Butte Daily Miner* of July 15 reported that, "...At 11:07, just one hour and 27 minutes from the time of leaving Butte, the first passenger cars over the Anaconda branch stopped at the lower end of Main street. The people of that flourishing place were taken by surprise and turned out in masse to learn what the arrival meant...In a few minutes the whole party repaired to the great smelter on the other side of the creek...a monument to the forethought and good judgement of the Hon. Marcus Daly, of the builder and superintendent, Mr. McCaskle and the pluck and enterprise of the owners, Messrs. Haggin & Tevis. Four hundred men are now employed there...

"A large town has been built near them and building is still going on. Mrs. Schultz of the Grand Central has enlarged and improved her hotel to metropolitan proportions and can easily accommodate 200 boarders. Cohen's new rink and hall building is fast approaching completion and in a few days a grand ball for the benefit of the school will be given in it. A large number of private residences and several substantial business houses have been built during this season. Anaconda is truly marvelous in its growth and stands in evidence of the faith, courage and enterprise of its citizens..." [2]

While the article mentions four hundred men being employed at the new smelter, there were 1,200 men on Daly's payroll and probably double that number working for the company on jobs related to the smelter operation supplying such necessities as bricks, metal products and timber. In addition, there were hundreds doing business with or employed by those doing business with the company. The Utah and Northern alone had three hundred men building the railroad spur from Stuart.

Also, Daly continued creating specialized companies with their own work forces, separate from his smelter, to provide services and materials which the smelter needed. It was about this time that he consolidated his supply and foundry operations and formed the Tuttle Manufacturing and Supply Company, in partnership with C.A. Tuttle, to manufacture iron and steel pieces and implements for use at the smelter. He also hired a young Englishman, George Perry, to build the furnaces for another new enterprise, the Standard Fire Brick Company, to supply brick for his furnaces and converters. Perry an experienced brickmaker, was associated with the enterprise the rest of his working life.[3]

The Haggin, Hearst, Tevis, Daly syndicate was pumping money into the operation at over one million dollars a year. Bob Vine reports that the cost of the Upper Works plant alone, exclusive of machinery, which started production in September, 1884, cost $1.6 million, and that operating expenses from September 23, 1884 to December 31, 1886 came to $3.2 million. Isaac Marcosson, in his official history of the Anaconda Company, says that the syndicate invested steadily in Anaconda for fifteen years without taking a dividend or dividing profits. Whatever was earned was immediately plowed back into the business.[4]

Not everyone was happy with this arrangement, however. Lloyd Tevis, early on, raised questions about the wisdom of making such extensive investments in copper and the new smelter. One can imagine Daly's fiery reaction to this questioning of his judgement. He insisted that all of the partners come to the site of his new smelter and see for themselves what he was doing. They arrived in July 1883 and stayed at the company headquarters at the Vineyard ranch. Haggin, a silent and dower man, made the trip under protest and seemed to liven up only when shown the trout that were being caught in nearby Warm Springs Creek. But he backed Daly all the way and resisted Tevis proposal to cut back on operations or at least to have Daly finance his own quarter interest in the enterprise. His message to Daly was, "When you need money, draw and keep on drawing." His answer to Tevis was, "I will see Daly through."[5]

Daly was a product of the California and Nevada booms, where fortunes were made and lost overnight, where the stakes were high and everyone played for keeps. He had worked as a mine foreman for the legendary John W. Mackay. Daly, much like Mackay, was a jovial, tobacco chewing, outgoing man of simple tastes and with a certain compassion for the downtrodden. But also like Mackay, he was dead serious and single-minded about mining. It was a gamble and you had to ante up if you expected to win. But when you had money, you went first class, ordered only the best of everything. Mining camps throughout the West were notorious for this spirit. Fancy foods such as oysters, fresh fish and game and foreign delicacies were common. The missionary Episcopal Bishop Tuttle in writing about Helena, Montana in 1868, observed that local ladies coming to call on his recently arrived wife arrived "arrayed in silk and adorned with gold and jewels".[6]

Daly, like his principal, Haggin, was openhanded with money. He did not believe in skimping or cutting corners. He had faith in his enterprise and was determined to make it the biggest and the best. His reaction to any problem or setback was to expand, to invest more, and Haggin was always there giving his support.

The first serious test of this operating philosophy came as the surge of copper production from the new smelter began to impact prices on the world market. When the Upper Works began operating in October 1884, with one concentrator, hand roasters and two reverbatory furnaces, it was declared by the *Mining and Scientific Press* to be a "wonder" with the largest concentrator in America, a mammoth smelter capable of treating 450 to 500 tons of ore daily. In fact, it processed 500 tons of ore from Butte every day from its inauguration in 1884 through 1885. Even as prices softened in 1885, Daly continued production at full capacity and raised havoc in the market by selling 7,000 tons of refined copper in England at greatly reduced prices.[7]

Because the Butte ores had a large silver content, Daly could make up part of his loss on copper with silver. In fact, starting in 1883, with the exception of Colorado, Montana was the largest silver producer in the country, much of it coming from the Walker Brothers Alice mine, the French-owned Lexington, the British-owned Blue Bird and W.A. Clark's numerous properties, all on Butte hill. By 1887 Butte was number one in silver.

However, with regard to copper, in the spring of 1886, Calument and Hecla, agent for the Lake Superior mines, probably in retaliation, cut its prices from 15 1/2 cents a pound to 10 cents. This move forced many smaller mines to close, but Daly stood firm, continuing full production, with a full work force at full wages.[8]

Daly's response was to look for ways to reduce production costs and thus improve Anaconda's competitive position. He sent his superintendent, Otto Stalmann, to Europe to bring back the latest technology. Stalmann, recognizing that times were tough, indicated to Daly that he could probably make the trip for $2500. Daly's reply, which was to become legend, illustrates the Daly, Haggin business philosophy. Handing him a check for $10,000, he said, "If you go to Europe for Anaconda, bear in mind, you go as a gentleman. When this money runs low, draw more." [9]

In 1886, Stalmann, in order to increase production and lower costs, began replacing the crusher and rolls in the concentrator with steam stamps and introducing Bruckner roasting furnaces. This would double the capacity of the Upper Works to one thousand tons a day. But by August the market was so bad that the company had to close down completely, in spite of Daly's expansion plans, throwing thousands in Butte and Anaconda out of work. This was the first of what came to be an all-too-familiar boom and bust cycle for residents of the new town of Anaconda.[10]

But this wasn't the first hardship the new town had experienced. During the previous winter, a fire had started in Jim Hamilton's store on the west side of Main Street between Front and Commercial and spread along the entire block. Even though there was Pat Maloney's volunteer fire department, there was no running water nor any fire hydrants in the town. So the blaze was fought with the only thing available, snow and snow balls. While most of the block was destroyed, the snow-ball brigade managed to contain the fire within that single block area.[11]

The winter that followed the closing of the smelter brought not only unemployment but exceptionally cold weather, with blizzard after blizzard piling snow on deserted streets and cold furnaces. This was the legendary cold winter of '87 that wiped out the huge cattle ranches in Eastern Montana, and made cowboy artist Charlie Russell famous for his sketch of a single scrawny cow frozen upright against a raging blizzard and wolves lurking hungrily nearby, and its descriptive title, "The Last of 5000". The only activity most of that winter at the smelter was that of a small work force, completing the installation of the new concentrating and smelting equipment.

By January 1887, prices were firming sufficiently that the Company readied itself to resume operations and by year's end ore production out of Butte had surpassed that of Lake Superior making it the nations number one source of copper.

What happened next should have convinced even the most doubtful skeptics that Daly and Haggin had infinite faith in the future of copper and their own operations and to believe Daly's assertion that his town of Anaconda was not just another mining camp offshoot on the way to becoming a ghost town. The Upper Works, even with the new equipment, didn't have the capacity to handle all the ore coming out of Butte hill. So Daly pressed ahead with plans and even larger investments for a new smelter located about a mile further down Warm Springs Creek, which was immediately dubbed the Lower Works.

Investments in the new smelter were accompanied by growing investments in the town, which began changing from a temporary boomtown of single men, large boarding houses, gambling houses, and all-night saloons, and started taking on the trappings of a more settled community. The resumption of work brought more people to the town. The number of women and children was slowly increasing. Wives and sisters joined husbands and brothers to turn bachelors' quarters into more homey environments. Newly arriving females were overwhelmed with attentions. Matches were made, wedding bells rang, new churches and schools were opened. By November 1886, in addition to the Central school on Main and

Fourth Streets, two elementary schools had been built, the Prescot on Park and Elm Streets, with Miss Lizzie Evans, later to become Mrs. Hoge, as principal, and the Lincoln, a six room schoolhouse on Fifth and Chestnut, with Miss Thompson as principal. At that time there were 299 children enrolled in the three schools.[12]

Fraternal societies, such as the Odd Fellows, the Knights of Pythias, the Free Masons, The Ancient Order of Hibernians and the St. Peter and Paul Society, were formed and began planning new buildings for their activities. Two large halls were built for meetings and entertainment, Daly Hall and Evans Hall. But the numerous saloons continued to be the favorite gathering places for most of the men when they weren't at work.

The fire of 1885 prompted the installation of water mains and a sewer system. The volunteer fire department was expanded and the first hospital, organized by Drs. Mitchell and Snyder, began functioning on the corner of Third and Main streets and wooden sidewalks were completed in the commercial section of town.[13]

In connection with the construction of the Lower Works, it was decided to build a small residential section between the roaster furnaces and the converters to house the men who worked there. It was named Carroll after Mike Carroll, Daly's General Superintendent in Butte. It had 125 houses, two large boarding houses, a public school, a general store and post office. The two biggest-selling items at the store in this community of mostly single men were beer and jugs of whiskey. All of the houses in Carroll were painted red and the story goes that this was the result of a drunken initiation rite when all who attended pledged to paint the town red.[14]

The Lower Works was completed late in 1888. Very early on a wintery March morning in 1889, the wind blew a spark from the hoist boiler room into the rope house, and for the second time, fire took its toll on the fortunes of Anaconda. The flames spread quickly through the buildings of the new smelter, "which swelled to white heat and fell in a sickening crash," according to one witness. This was a major set-back for Daly and the syndicate, which had struggled with cash-flow problems to build it and had counted on the increased production to capture a major share of the growing world market.Haggin, with characteristic calm and determination wired from San Francisco to Daly, "Don't lose heart. When you have time, advise how badly we are hurt."[15]

2. Center section of Anaconda looking north about 1890. The Montana Hotel dominates Main Street. Next is the Leland Hotel, and across the street are the offices of the *Anaconda Standard*. Continuing south are the Presbyterian church and the Central School. Further along is St. Mark's Episcopal church and directly behind it on Sixth and Oak Streets is St. Ann's Hospital. In the far upper right note the steeple of St. Paul's Catholic Church on the corner of Park and Cherry Streets.

The rebuilding of the Lower Works began immediately and was finished in record time, with operations once more under way in September, 1889. This time the mills and smelter were constructed of the most modern of building materials, structural steel beams and sheet iron. Structural steel beams, which were being used to build the Washington Monument and had been used in the recently inaugurated Brooklyn Bridge, were revolutionizing the construction industry.

Daly's and Haggin's confidence in the future of Anaconda was enthusiastically seconded by a growing population and the continued construction of houses, halls and commercial buildings. Installation of water mains and sewers was proceeding apace.

Religious congregations, which had been holding services in public halls or in the two existing church buildings began the construction of their own churches. The Catholics, who had been conducting occasional services in the Odd Fellows Hall and Evans Opera House, whenever a priest arrived from Deer Lodge, built their first church on the corner of Park and Cherry streets and named it St. Paul's. Father Peter DeSiere, a young Belgian priest from the Deer Lodge mission, and George Barich had organized the new parish, directed fund-raising and the building of the church. Father DeSiere was named the first pastor. Regular services were inaugurated in September 1888, and it was formally blessed by Bishop Brondell of Helena on November 25. Eighty-two baptisms and twenty-two marriages were recorded for that year in the church records.[16]

The Presbyterians, who had been holding services in the Methodist South Church since 1884, held a meeting in February 1886, sponsored by two missionary ministers, the Reverend J. M. McMillan and Reverend E.P. Linnell to organize the First Presbyterian Church of Anaconda. The twenty men and women at the meeting became charter members. Mr. Donaldson Greer and Mr. Frank Hoagland were elected elders. The Montana Board of Home Missions appointed T.J. Lamont pastor of the new church. Less than three years later, on December 9, 1888, the first service was held in the new church on the corner of Main and East Third Street. At this same time, the Methodists were planning a new church to replace their original building, and plans were made for an Episcopalian church also to be located on Main Street.[17]

While the new churches were being built and planned, Daly was demonstrating in yet another way his confidence in the future of his town by building a new hotel, intended to be the best in the West. Montana was about to become a state and Daly was already preparing the way for his new town to become the state capital. As young as she was, Anaconda was Daly's queen, and he would dress her in the richest finery and place a scepter in her hand. The new hotel would be her finery and the new capitol her scepter.

The territorial capital, briefly established at Bannack and then Virginia city, had been moved to Helena when Last Chance Gulch became the center of gold mining in the territory. With copper becoming the metal of the future, what better place for the new capital city than the town that copper was building.

Daly's partner in the bank and about to become the city's first mayor, William Hoge, became a share holder and vice president in the new association formed to build the hotel. They hired W. W. Boyington of Chicago to do the architectural plans and D. F. McDevitt of Butte as supervising architect and builder. Daly would have nothing but the best for his new Montana Hotel. The terra cotta trim and decoration was shipped in by train from Chicago, the bar, a reproduction from the famous Hoffman House, unofficial headquarters and watering hole of the Democratic party in New York City, was made of solid mahogany, as was all the woodwork in the bar room. The original plan for the hotel, to be located on the south west corner of Park and Main streets, was for three stories, but in October, 1888, Haggin, on one of his infrequent visits to Anaconda, saw the structure nearing completion and suggested an additional story.[18]

The hotel was built in the form of a U so that all the rooms would have outside light. When completed, the building took up one quarter of the block with its principal entrance on Main street. The entire ground floor was devoted to public rooms such as the bar, restaurant, breakfast room, a chandelliered ball room, club rooms, billiard room and barber shop. The space between the two legs of the U was roofed over at the ground floor level providing an ample reception area and lobby.

It was equipped with the latest and most modern equipment, including telephones, both gaslight and electric lights, steam heat, and the sensation of its day, an electric elevator to whisk guests four stories high. Many of the rooms had canopied beds, imported Italian marble fireplaces and separate bathrooms, and many of the public rooms were carpeted with Persian rugs. William Hoge had traveled to the east coast to personally select furnishings. While in St. Paul, Minnesota, he hired D.L. Harbaugh to be the hotel manager and recruited a staff of fifteen negroes from New York to serve as waiters, porters and janitors. They would thus become the founding members of the town's small colored community.[19]

With the opening of his new 185-room Montana Hotel, Daly announced to the world that his town, Anaconda, barely five years old, was a permanent and major factor in the new state and would become one of its most important cities. To make sure that no one missed the point, even while the Lower Works were still being rebuilt, Daly held a grand opening ball on July 4, 1889, sending out 1500 invitations. Special trains ran from Helena, Deer Lodge, Great Falls, Butte and Dillon to bring guests to the hotel. A troupe of Shakespearean actors, presenting *As You Like It* at Evans Opera House, was persuaded to stay over to attend the inauguration.[20]

The hotel became not only the social and physical center of Anaconda, but its symbol as well. It was the town's first really significant building, erected before either the City Hall or the Court House. It gave Anaconda a prominence and class that most other cities in Montana lacked. From that time forward, the fortunes of this structure would reflect those of the city itself. It stood proudly through good times and bad for almost ninety years but was finally gutted and decapitated in 1978. Its demise preceded by two years the closing of the smelter thus forecasting the end of an era, if not of Daly's town itself.

Other building around the new town was proceeding apace even as the hotel went up. The Snyder hospital had been destroyed by fire and a new hospital was being built by Dr. Leiser on the corner of Sixth and Oak streets. Shortly after it was completed, the Sisters of Charity of Levenworth Kansas, who were operating hospitals in Deer Lodge and Helena, were persuaded to purchase the one in Anaconda as well. It opened it doors to the public on July 26, St. Anne's day, 1889, and was named in her honor. This eventually became the city hospital, with the Company deducting $1.00 a month from each employee's wages for medical services at the hospital.[21]

Local elections were first held in 1888. William L. Hoge, Marcus Daly's partner in the bank and hotel, became the first Mayor. The city council representatives were Jack Morris and W. Sutter for the first ward; A. L. Kempland and A. C. McCallum, second ward;

3. The Montana Hotel. The town's most significant structure. Built in 1889 before either the City Hall or Court House, it gave Anaconda both prominence and class and became a symbol of its well-being through good times and bad.

Joseph Peters and I.F. Kirby, third ward; H. T. Mahan and J. C. Keppler fourth ward. James McKirety was elected city treasurer, F. D. Fitzgerald, police magistrate and George B. Winston, city attorney. The first city council meeting was held in the office of F.D. Fitzgerald on August 25, 1888.[22]

This and subsequent meetings passed the first ordinances governing the city. They included such matters as dividing the city into wards for political purposes, confirming the appointments of the Chief of Police and Fire Chief, restricting the use of firearms, prohibiting the carrying of concealed weapons, declaring habitual drunkenness against the law, banning hurdy-gurdy houses and forbidding the employment of females as waitresses, bartenders, servants, singers, dancers or musicians in saloons or other places where liquor was sold. Fees and licenses were established for wheels of fortune, street musicians, astrologers, seers, fortune tellers, and knifeboard owners.

One of the first petitions presented to the city council was from Daly's accountant, William Read, who requested that Chinese laundries be regulated. By this time, Anaconda had attracted a population of at least 300 Chinese, most of them living on both sides of one block on Birch Street between First and Second, later to be called Commercial and Park. The City Council also entered into contracts with the Anaconda Company to furnish the city with fire hydrants at $70.00 a year and twelve all-night electric lights for $16.00 a month. The street lights were inaugurated on Christmas eve 1888, with due note being taken that Anaconda was one of the few towns in the country with electric street lights.[23]

The following year a city park, with abundant shade trees, running brooks and small lakes and ponds was created on the banks of Warm Springs Creek near the Upper Works. Here picnics could be held and informal games and athletic competitions could take place. Matt Kelly reports that on June 27, 1889, the United Irish Societies of Butte and Anaconda held a picnic there, with an estimated four thousand people in attendance. Fourteen coaches arrived from Butte and stopped at the east end of town so that the Butte societies of the Ancient Order of Hibernians and The Robert Emmett Guards of Butte and Walkerville could parade through town to the park on the west side. The park also became a convenient overnight camp for wagon trains passing through.[24]

5. The Bad Medicine Wagon

If gold had been the enticement for exploring the West, the railroad was the force that impelled its settlement and transformation. Since the new town of Anaconda was a product of that transformation, let's take a quick look at the impact railroads had on developments in that part of Montana.

Originally part of President Jefferson's 1803 Louisiana Purchase and successively included in the Territories of Missouri, Nebraska, Dakota and Idaho, Montana first became a separate Territory in 1864. For six decades after the Lewis and Clark expedition, it remained a vast wilderness of buffalo, Indians, fur trappers, US Army outposts, missionaries and a few brave settlers in sod huts or log cabins raising cattle or trying to eke a living from the soil.

Until the Union Pacific railroad made its way across Nebraska and Wyoming and became the first transcontinental line by joining up with the Central Pacific in Utah in 1869, the primary routes of access into Montana were a 2,500 mile journey by steamboat up the Missouri river from St. Louis to Fort Benton or by wagon train along the Mormon, Oregon and Bozeman trails. At the time of the completion of the Mullan military road from Fort Benton to the Columbia River at Walla Walla in 1860, records showed only twenty white men living in the territory. Even after a decade of gold and silver fever, the census of 1870 indicated a population of only 29,595. This census also records the interesting fact that the number of people remaining in Butte in 1870 after the placers played out was 241, ninety eight of whom were Chinese.[1]

But the influx of white settlers and miners into the West resulted in increasing conflict with the native population. The invasion of Indian hunting grounds by a growing surge of homesteaders and gold seekers touched off a series of Indian wars. In 1862, the Sioux of the Dakota region went on the warpath and massacred or imprisoned almost a thousand white men, women and children. The Cheyenne, banished from their hunting grounds to the desolate wastes of southeastern Colorado began attacking the mining settlements to the north. They, along with the Sioux, began crowding into eastern and southern Montana.[2]

By 1868 the Montana frontier was experiencing serious Indian trouble. So the army increased its presence in the area and by 1892 had established a dozen forts in the Territory. In 1863, John M. Bozeman had blazed a road to the Montana gold camps, which departed from the Oregon Trail west of Fort Laramie crossing the Big Horn River and following the Yellowstone valley west. Since it invaded Indian hunting grounds in the Big Horn Mountains, its use set off prolonged and bloody confrontations with Cheyenne and Sioux all along its route. Wagon trains using it were regularly attacked and Bozeman, himself, was killed by Piegans on Mission Creek, east of present day Livingston, in 1867. Further north, the Blackfeet, Blood and Piegan also attacked settlers traveling the Minnesota-Montana overland route. These attacks grew so fierce, that in 1868, the U.S. Government withdrew its troops from Fort C.F. Smith, which it had established two years before to protect the Bozeman Trail, and declared the region a reservation.[3]

The coming of the railroad only exacerbated an already tense situation. The Indians saw the iron horse as a direct threat to themselves and their way of life. Tribes which had been removed from areas further east and had accepted government guarantees of land in perpetuity, now faced an invasion of what they called, "the bad medicine wagon". At stake were not only the treaty lands but the large herds of buffalo which roamed freely over them. Both the material and spiritual well being of the Plains Indians was based on the buffalo. It not only supplied food, shelter and clothing but was a mythical hero and religious symbol.

The guns of the Union army had scarcely fallen silent after Lee's surrender at Appomattox Courthouse, when they began belching fire again, this time for the railroads and against the Indians. Union generals, whose names had become household words during the Civil war, once again gained prominence in reports of skirmishes with Indian tribes in Kansas, Nebraska, Colorado, the Dakotas, Wyoming, Montana, Idaho, Utah and California.

Completing a transcontinental railway was made national policy by the Pacific railway act of 1862. But even earlier, in 1853, Isaac I Stevens, who was later to become the first Governor of the Washington Territory, surveyed a route for a railroad which would cross the present state of Montana on its way to Puget Sound. The Railway act provided to the railroads out of public domain blocks of land ten miles wide on alternate sides of the rail right of way for each mile of track as the road advanced. This amounted to 6,400 acres per mile of track, the government keeping the alternate blocks. In addition, for each forty-mile section completed, the railroads would receive United States bonds amounting to $16,000 per mile on the plains, $32,000 per mile on the plateau between the Rockies and the Sierra Nevada and $64,000 per mile in the mountains. This was considered a loan to help pay for construction and was to be repaid out of company profits once the railroad began operations.

From Sacramento, California, the Central Pacific was organized to lay rails east. Two railway groups, the Union Pacific and the Kansas Pacific, competed to become the first line to reach the 100th meridian and thus be designated to finish the line west. In 1864, even before construction began on the first transcontinental railroad, President Abraham Lincoln chartered the Northern Pacific Railroad Company to build an additional road from Lake Superior to Puget Sound which would utilize part of the route that Stevens had surveyed a decade before. The land grant given this group was even larger, 12,800 acres per mile across states and 25,600 across territories. In total this grant gave the Northern Pacific 47 million acres, more land than was in the entire state of Connecticut.[4]

Even though the treaty lands had been deeded in perpetuity to the Indian tribes that occupied them and the Supreme Court had ruled specifically that railroad land grants did not apply to lands "which Indians, pursuant to treaty stipulation, were left free to occupy", the railroads through bribery and manipulation maneuvered the Interior Department's Land Office and the Congress to "extinguish" any and all Indian claims which got in their way. In some cases the railroads tried to make the Indian cession of lands legal by drawing up contracts and making token payments to the various tribes.

We are given a rare and penetrating insight into this practice by the following excerpt from the transcript of the Council meeting held by Joseph K. McCammon, Assistant Attorney-General, appointed by the Secretary of the Interior to negotiate with the Indians on the Flathead Reservation in Montana for right of way through their lands for the Northern Pacific Railroad:

August 31, 1882

Indian leaders present: Arlee, Adolphe, Eneas, and Michelle, with headmen and Indians of the Flathead, Pend d'Oreilles, and Kootenais tribes.

Commissioner McCammon: " My friends of the Flathead, Pend d'Oreilles, Kootenais, and other tribes living on the Jocko Reservation, I have been sent by the Great Father at Washington a great many miles to see you and talk with you. He knows how well you have treated the white man all these many years;...Twenty-seven years ago you and your fathers made a treaty with the whites....In that treaty you and your fathers agreed 'If necessary for the public convenience roads may be run through said reservation.' The Great Father and the Great Council in 1864 gave the Northern Pacific Railroad Company the right to build a railroad through this country....

Eneas: "I presume you will not ask us to answer now. There are some men here who have wild ideas, and we want to adjourn and talk the matter over...

September 1, 1882

Commissioner McCammon: "My friends, I am glad to see you today, and hope your hearts are good towards the Great Father..."

Eneas: "I am the chief and you see me now, I have no doubt you are sent to see us by the Great Father. I am the chief and this is my country. I am not joking in telling you I would like to get the Flathead Lake country back. There are things that the government promised me in that treaty that I have never seen... The government told me it would send a blacksmith, and build school-houses, and furnish teachers...and a head farmer and build houses for us. I was glad to think we were to have these things. We had a big country, and under those conditions we signed the treaty. Seven years after that we learned that the line of the reservation ran across the middle of Flathead Lake. We didn't know that when we signed the treaty. That is the reason we want that country back ...We are poor now. We try to have whites to assist us, and they won't because we are Indians..."

Commissioner: "I have not power to treat about everything."

Eneas: "...My country was like a flower and I gave you its best part. What I gave I don't look for back, and I never have asked for it back. The Great Father gave it to us for three tribes... What are we going to do when they build the road? We have no place to go.

That is why it is my wish that you should go down the Missoula River. I am not telling you that you are mean, but this is a small country and we are hanging on to it like a child on to a piece of candy."

Commissioner: "The line selected by the railroad company was selected ten years ago, because it was the best route and because down the Missoula River would not be a good route...The Great Father will be sorry when he hears that the Indians do not believe his good faith...and if I go back without your having named a price for the lands, he will say they are not the good Indians and faithful friends I thought."

Arlee: " We don't think anything bad, but we don't want the railroad to go through the reservation here, because these white men are bad people...because when the white men come into work there will be trouble, that is all... It was nine years ago that Garfield said 'Don't think we will thrust you from that country; that land belongs to you.' Last winter, I was at home lying down when they told me men were surveying the place. Some said it did not amount to anything, but I said it would cut our reservation in two; and now today I see you here trying to get our land from us for the railroad. But I do not want any railroad here, for this is my country."

Commissioner: " This is your country; there is no doubt about that... I will again ask you if you can name what money you want for this right of way. If you cannot, I will name a sum for you... The Great Father told me to propose a fair price, and I think that $10 per acre is a fair price for 1,500 acres. That is four times as much as the Great Father gets for his land. This would make altogether for the land $15,000. The Crows got only about five thousand, and the Shoshones seven thousand, or nearly eight. In addition, each Indian will be paid for his fences and barns where this railroad interferes with him. The $15,000 will be for the benefit of the whole tribe."

Arlee: " I want $1 million for it."

Commissioner: " The whole reservation would not be worth that."

Arlee: "I thought you were here to help us."

Michelle: "When I heard you the first time I was glad; but now when I hear what you offer, I do not feel so well, because now you say that all the reservation is not worth $1 million. Now I do not agree with you... When the railroad runs through the railroad company will get the money back in one day. They will run through my ranch and take my timber to build it with. I would not take $15,000. I do not mean we will make trouble; I only say we will not take $15,000. I do not speak now, any more, because you offer only $15,000."

Commissioner: "We will not talk any more about the million dollars; the Great Father will not allow us to talk about that."

Arlee: "All right; then go by the Missoula. If the railroad don't want to give the money, let it go by Frenchtown."

After another meeting the next day the Commissioner raised the price to $16,000. When Arlee asked for the money in cash, he got this answer from McCammon: "...The money will be expended for the benefit of the Indians in the manner the Great Father thinks best. If he thinks, after hearing from you, that it is better to let you have the money, he will pay the money. You must depend upon his judgment as to how the money will be paid. The Great Father will never forget you."[5]

While this exchange took place in 1882, it is an eerie repetition, almost to the exact words, of similar meetings in the West between U.S. government representatives and Indian Chiefs going back to the 1850s.

As the Union Pacific moved west in the summer of 1866, Cheyenne and Sioux continually attacked survey parties, tie cutting and grading crews. They killed and scalped six members of a section gang 220 miles west of Omaha, chased a survey party at Ogalala, pulling up their stakes, killed an assistant surveyor and a trooper at Rock Creek, burned stagecoach stations and stole livestock at Powder River. In December of the same year, they ambushed and killed eighty-one soldiers. Railroad men, stage agents, ranchers and settlers all clamored for more protection.

In 1867, the Kansas Pacific railroad hired a twenty-one year old marksman named William Cody to supply buffalo meat for the workmen on the construction gangs. Thus began a process, repeated by other railroads and joined by indiscriminate hunters, which would eliminate the buffalo from the western plains by 1883. As the Kansas Pacific pushed west from Abeline, the Indians fought back, challenging surveyors, graders, section gangs, and train crews. In the spring of 1867, hunting parties of Cheyenne, Sioux and Arapaho roamed west of Salina chasing off section gangs, ripping up track and robbing or destroying railroad supplies.[6]

Many of the new railroads also had retired Union officers on their boards of directors or in prominent leadership positions. Thus, they could get a sympathetic ear from former fellow officers still in uniform. Ex-Brigadier General William F. Palmer, still in his twenties, was treasurer and director of construction on the Kansas Pacific and became the driving force in raising money and completing its construction to Denver. The Union Pacific hired General Grenville Dodge as chief engineer and General Jack Casement as construction supervisor, after which, the work on that road took on the appearance of a military operation.[7]

In reaction to the attacks in Kansas, General William T. Sherman, in command of military forces in the West, devised a plan to drive all the Plains Indians north of the Platte and south of the Arkansas River leaving a broad belt for transcontinental railroads. During the summer of 1867, General Winfield Scott Hancock with Major George Armstrong Custer of the Seventh Cavalry chased Indians across Kansas, burning Indian tepee villages and killing indiscriminately, signaling the beginning of a long and bloody Indian war. Troops actively patrolled the railroad lines and General Philip Sheridan, believing that the only good Indian was a dead Indian, gave orders to track down and kill any and all Indians sighted near the railroad. By the summer of 1869, Indian resistance along the Kansas Pacific had been smothered, but the fight went on further north, along the Union Pacific and then the Northern Pacific.[8]

The first railroad into Montana Territory came from the south, the Utah and Northern, a branch of the Union Pacific from Corrine, Utah, which crossed into Montana in 1880, but was stopped near the abandoned mining camp of Bannack for over eight months while the railroad negotiated a right-of-way with a recalcitrant land owner. The railhead grew into a small community and was named Dillon after Sidney Dillon, President of the Union Pacific Railroad. It would later became the seat of Beaverhead county. In December 1881, the railroad crossed the continental divide into Butte. Thus, the ore that had been previously carried in oxcarts to Corrine, would now go by train.[9]

In 1871, Indian Chief Sitting Bull, who had kept his distance from the white man and his railroads, always retreating ahead of settlement, felt that his refuge in the Big Horn mountains was being threatened when Northern Pacific surveyors were discovered running a line through the Yellowstone valley. The Indians drove them out and when they returned the following year drove them out again.

In 1873, at the request of the owners of the Northern Pacific, the War Department organized a powerful military force of more than 1500 officers and men, 353 civilian mule drivers guides and interpreters, 275 wagons and ambulances and 2300 horses and mules to serve as a protective escort for an engineering survey party in the valley of the Yellowstone. The force was commanded by General David Stanley and Lieutenant-Colonel George A. Custer. During July and August, Indians, under Sitting Bull and Crazy Horse, engaged Stanley and Custer in a number of indecisive skirmishes, but which in retrospect can be read as forerunners of Custer's disastrous defeat at the hands of Sitting Bull on the Little Big Horn in eastern Montana on June 25, 1876. It was only six weeks later that Marcus Daly entered western Montana by stage coach from Salt Lake City and first set foot in the silver camp of Butte.[10]

After the Custer battle, in the fall of the same year, Sitting Bull with a large band of warriors attacked a pack train of supplies bound for the Fifth Infantry under Colonel Nelson A. Miles at the mouth of the Tongue River. After hot pursuit by Miles, Sitting Bull broke away from the main body of his force and headed for Canada. Miles then initiated an aggressive winter campaign which culminated in the surrender of Crazy Horse with over eight hundred of his people and two thousand horses. In March over two thousand more Indians surrendered. By spring only Sitting Bull was still at large.[11]

But it wasn't the Indian wars that slowed down the building of the railroads, it was economic difficulties. Early in 1873, the Union Pacific was tarred with the brush of the Credit Mobilier scandal which exposed the graft and corruption involved in the UP construction. Then in September, the banking company financing Northern Pacific construction declared bankruptcy. The two events brought on a serious depression, the famous Panic of Seventy Three. By that time the NP had reached Bismark, North Dakota. It was not to advance beyond that point for seven years. Finally in July 1881, the company, under new management and with new financing, reached Glendive, Montana territory. Construction was halted for the winter at what later became Miles City, named after Nelson Miles, recently promoted to General and commander of nearby Fort Keogh.

The following summer, work proceeded rapidly to meet crews working eastward through Idaho and western Montana. By June 1883, rails had reached Helena, the territorial capital. The last 300-mile section lay in the mountains and required the construction of tunnels under the long Bozeman Pass and further west the Mullan Pass at elevations exceeding 5500 feet. Butte miners were hired to finish construction on the last portion of the Mullan Pass tunnel because of the need for their expert timbering skills.[12]

Just three months after Marcus Daly had founded his new town on the banks of Warm Springs Creek, on the headwaters of the Deer Lodge River, another historic act, important to Anaconda's future, took place further down the river at Gold Creek. On September 8, 1883, a ceremony was held to celebrate the completion of the Northern Pacific railroad. It was here that Henry Villard, the financial wizard who had made it possible, drove in the last spike on the nation's second transcontinental railroad. In attendance for this ceremony were 350 guests who had been transported from St. Paul, Minnesota in four luxurious trains of Pullman cars draped in patriotic bunting. Among the guests that day were former President U.S. Grant, General Philip Sheridan, Secretary of the Interior Carl Shurz, British statesman and historian James Bryce and, irony of ironies, Chief Sitting Bull and members of the Crow tribe, which had ceded large swatches of its treaty lands to the Northern Pacific.[13]

The same year, the Union Pacific, expecting competition from the Northern Pacific, had extended its own Utah and Northern line northwest past Butte through Silver Bow Canyon and along the Deer Lodge river to Garrison, within spitting distance of Gold Creek. It was from this line that the Utah and Northern built a branch from Stuart to Anaconda in 1884. On the day of Villard's inauguration of the new line, the Northern Pacific had arranged with the Utah and Northern to nail benches on flatcars and offer free transportation to all residents of Butte and Anaconda who wanted to attend the ceremonies.[14]

In 1886, UP and NP agreed on a plan for a joint standard gauge line between Garrison and Butte. The new company was named the Montana Union. As part of this agreement the UP was given trackage rights over the NP line from Garrison to Helena. The railroads saw enormous profits to be made transporting ore and refined metals out of Butte and Anaconda and fought desperately to maintain a dominant position with the mine owners. The UP, being first on the scene, had locked up the primary traffic and, now, by forming a joint road with the NP, had effectively eliminated any competition.

Anaconda Copper was in a fight for markets and needed lower rates and dependable service to move its rapidly expanding production. It got neither from the Montana Union. In the exceptionally cold winter of 1886-87, one of the reasons Marcus Daly decided to shut down his operations was because the Montana Union failed to deliver the coal he had contracted for.[15]

Thus, it was with some satisfaction that Anaconda received the good news in 1887 that James J. Hill in one blazing summer had laid 643 miles of track on his Great Northern railroad from Dakota Territory to Helena. The following year he would extend the line south into Butte. Hill promised his friend Daly "all the transportation you want at rates as will enable you to largely increase your business. What we want over our low grades is heavy tonnage, and the heavier it is, *the lower we can make the rates."* (Hill's italics)[16]

But the railroads offered Haggin and Daly more than just a way to get their copper to market. They also provided an ever growing pool of abundant and cheap labor for the mines and smelters. In addition to thousands of Irishmen hired to build the railroads west and already making up the majority workers in Butte and Anaconda, all of the railroads had agents in Europe recruiting new immigrants to populate and make productive the vast acreages that they had been granted. In the early 1880's the Northern Pacific had 831 agents in Great Britain and 124 scattered across Norway, Sweden, Denmark, Holland, Germany and Switzerland. The railroads also actively competed with each other for groups of immigrants that had already arrived in the United States. Omaha, one of the gateways to the west, was filled with agents selling railroad land to new arrivals, diverting them from their intended destinations with tales of Indian raids and fearful hardships, employing religious leaders to attract settlers and offering bribes to immigration officials to route new migrants to them.

All of the railroads painted bright pictures of the richness of the soil, the abundance of rain and the benign climates of the regions through which their roads passed. This was big business. The railroads needed to convert their land into cash and needed the freight that production from that land could produce. Although by European standards land prices were reasonable, the railroads generated increased income through the excessive interest they charged on short-term mortgage loans. They also earned good profits from the passenger fares of the westward-bound immigrants. By the early 1880's, the Northern Pacific and other roads were running trains with as many as fifty cars filled with immigrants and their baggage. Special immigrant cars were built to accommodate large numbers in limited space. In the West these were called Zulu cars.[17]

6. *"Help, Help, Come Running!"*

Not all the Indian wars in the West were related to railroad construction. The one that early settlers in the Deer Lodge Valley remembered and that became part of the area's history was the Battle of the Big Hole. This was one of a number of battles of the Nez Perces war that took place in 1877 and held the attention of the entire nation.

It began over white settlement on traditional Indian land of Chief Joseph and his band of Nez Perces in eastern Oregon. As a result of the gold discovered on the Nez Perces reservation in Idaho in the 1860s, the treaty of 1855 was revised in 1863 reducing the reservation to one-tenth its original size, leaving a number of the Nez Perces bands on land outside the new reservation. These bands refused to sign the treaty and continued to claim their traditional living and hunting grounds. To them the land that they had lived on "from the beginning of time" was sacred. The earth was their mother, they were made from the clay of that earth and to that earth they would return to be born again. While they lived, the great spirit commanded them to protect that ground and their ancestors buried within it.

They held out against the encroachment of white settlers until 1877, when they were ordered off their lands and into the reduced reservation. Joseph and his band came from the Wallowa region of northeastern Oregon where the tribe had lived as long as they could remember. The bitterness generated by this order caused a couple of hot-headed young braves from another band in Idaho to take the various tribes' fate into their own hands by searching out and killing known Indian haters among the white settlers. Joseph, well-known among the whites for his decency and dignity, had successfully managed earlier confrontations where whites had killed Indians by demanding justice but restraining his followers from rash acts of vengeance when those responsible went unpunished.

Before the murders, he reluctantly concluded that the tribes could not resist the whites and had convinced his band to move from their traditional home. After the murders, however, the situation changed. All of the non-treaty tribes were inflamed. They had lost all faith in white-man's justice and believed that they would have to fight for their land and their rights. There was no turning back.[1]

In armed encounters with detachments of government troops and settlers, the Indians exacted a heavy toll, suffering few casualties themselves. The Army lost one officer and thirty-three enlisted men in its first and most disastrous battle at White Bird Canyon on June 17. Minor skirmishes followed into the middle of July with the Indians managing each time to avoid being surrounded and forced to surrender.[2]

As more troops were called in and more settlers joined the fight, the Indians took their families, belongings and their herds of horses and moved out of reach. With the U.S. Army under General Oliver O. Howard, several days behind them, they decided to leave the reservation area and head for Montana over the Lolo trail into the Bitter Root Valley. Led by Chief Looking Glass, who had fought with the Crows in Eastern Montana, they intended to seek refuge with the Crows and Sitting Bull on the Little Big Horn, unaware that General Miles had already forced Sitting Bull into Canada.[3]

Ever since the Indians began using horses in the late 1600's, the tribes on the west side of the continental divide would make the annual trek to hunt buffalo on the eastern plains. The Nez Perces and the Flatheads of western Montana often made this trek together and, just as often, would join to fight the Blackfeet who resisted the western tribes hunting east of the mountains. The two tribes were hunting partners and allies and there was considerable intermarriage among them. The Lolo trail was one of a number that the Nez Perces used to and from their annual buffalo hunt.

The new settlers in the Deer Lodge Valley had grown used to seeing these Indians pass through the valley and had even come to know some of them. It is said that Morgan Evans had dealt with Chief Joseph, who occasionally had bought flour at his grist mill. It is more probable however that it was Looking Glass, chief of a band settled on the Clearwater River in Idaho or Joseph's son Ollokot that he had come to know, because the Old Chief Joseph died in 1871. His other son, Hinmahtooyalatket, took over his mantle as chief and became known to the whites as Joseph. But the young Joseph, faced with the problem of white settlers on the tribe's traditional lands from the time of his father's death, preferred not to join the annual hunt. Ollokot, as chieftain of the fighting braves, went in his stead. There is no record that the young Joseph ever entered Deer Lodge Valley.[4]

This most recent Indian uprising, and the losses sustained by the Army, were reported throughout the United States. In light of Custer's defeat at the hands of Sitting Bull the year before, the army's apparent inability to contain the Indian problem in Idaho raised general alarm, especially in western Montana. The June 30, San Francisco *Chronicle* announced: "THE FLATHEADS PREPARING TO JOIN THE NEZ PERCES IN THEIR WAR ON WHITES". And, indeed, there was justifiable concern among the settlers in the Bitter Root Valley, traditional Flathead territory, and around the St. Ignatius Mission on the newly established Flathead Reservation.

A festering resentment against whites had been growing for years and became even more acute in 1871 when the government forcefully ejected the Flatheads from their traditional homeland and moved them into the St. Ignatius area. Two hundred warriors under Chief Charlo refused to go and continued to live and hunt in the Bitter Root. It was a tense time and only the long-time presence of Jesuit missions in the region helped maintain an uneasy peace. The priests counselled the Indians to avoid violence and look for alternative methods to defend their interests.

The Flatheads were a proud and fierce warrior people, whose bravery was legendary. Bishop James O'Conner of Omaha visiting the St. Ignatius Mission in the spring of 1877 noted: "The most warlike of all Rocky Mountain tribes, the piety of the Flatheads has not diminished their bravery, for since their conversion as well as before, they have been more than a match for their neighbors, the Sioux and the Blackfeet."[5] But the Jesuit presence was misunderstood and resented by many settlers and even some government agents and military men who accused the priests of opposing white interests.

When news of the trouble with the Nez Perces in Idaho reached the settlers in the Bitter Root, their immediate reaction was panic. Tension had been building for a long time. The year before, six months before the Custer massacre, Father Giorda, superior of the mission network, sent the following message to President Grant: "From information which I cannot disregard, I learn that the Flathead Indians in the Bitter Root, Montana, are restless and may be driven to the wall, go and join the hostile tribes on the plains." In December 1876, Flathead Indian Agent Medary wired General Gibbon at Fort Shaw for soldiers "at once". Gibbon could spare only twenty soldiers who stayed until March.

At first word of the Nez Perces trouble, most settlers in the valley abandoned their homes and fled to the crumbling adobe walls of the old trading post of Fort Owen at Stevensville. Rumors ran rampant. The Nez Perces and the Flatheads, traditional allies, would jointly invade the valley. The Sioux and the Crow had entered into an alliance to get rid of the whites. Nez Perces agents were among the Crow. Had any of these things occurred, the settlers in the Bitter Root would have had little chance. James Mills, secretary to Montana Territorial Governor Potts, estimated that there were approximately 500 braves among the Flathead, Pend d'Oreilles and Kutenais who might join the Nez Perces. This was more than a match for all the settlers and soldiers in the area. Among the settlers gathered at Fort Owen there were only ten effective rifles. Governor Potts, issued a call for volunteers. Emphasizing the emergency, the *Weekly Missoulan* ran a headline of "HELP, HELP, COME RUNNING," over Potts' declaration.[6]

Local officials working with the Jesuits took immediate steps among the Flatheads to obtain pledges of neutrality and even sought to raise volunteer forces from among them to defend against Nez Perces on the warpath. After getting pledges of neutrality from the Chiefs of the upper valley, Captain Charles C. Rawn from Fort Missoula and Indian Agent Peter Ronan, with the help of Jesuits Father Jerome D'Aste and Anthony Ravalli, met with Chief Charlo. Charlo declared that, in spite of the suspicions of the settlers, he was not ready to go to war against them. He reminded them, however, that the Nez Perces were allies and that he could not ask his young braves who had fought alongside them to now go to war against them. He would maintain strict neutrality, but would cooperate with the settlers by providing information. The Bitter Root Valley was Flathead territory and if the invaders molested his people, either red or white, he would fight them.[7]

Charlo resisted subsequent entreaties from Nez Perces Chief Looking Glass to renege on this commitment and insisted that the Nez Perces respect and maintain peace in the valley. The settlers later credited the guiding hand and calming influence of Father Ravalli on

Charlo and his braves for keeping the peace in the valley that summer.[8]

Since it was not clear where the Indians were headed, volunteer forces were raised in Missoula, Helena, Deer Lodge and Butte, as well as in the Bitter Root Valley. Also, General Philip Sheridan, in charge of the Army's Missouri district, dispatched General John Gibbon, in command of the Seventh Cavalry at Fort Shaw on the Sun River east of the Rockies, to undertake a forced march and intercept the Nez Perces and assist General Howard in returning them to Idaho.

The Nez Perces, on the other hand, expecting support from the Flatheads, hoped that they could make a transit either north to Canada or to Crow country by traveling through sparsely settled areas. So with Chief Looking Glass in charge, they started up the Lolo trail with two hundred braves, five hundred and fifty women and children, two thousand horses and all their worldly possessions.[9]

They met their first challenge on July 25 when they neared the Bitter Root exit of the Lolo trail. Captain Rawn, from a small forward detachment under Gibbon's command at the newly established Fort Missoula, had constructed a hasty barricade across the trail and ordered that the Indians not proceed further. When the Indians proclaimed their peaceful intent, Rawn demanded that they surrender their arms. When they refused, Rawn recognizing that he was greatly outnumbered even with the assistance of approximately one hundred volunteers from the Bitter Root Valley, attempted to stall for time expecting additional support from Gibbon or from General Howard who was still in pursuit from the west. Instead of attacking the greatly outgunned Rawn and his men, the Nez Perces, perhaps realizing they were in Flathead territory but did not have Charlo's direct support, merely skirted the barricade a couple of miles up the trail and headed up the Bitter Root River. [10]

The volunteers, hearing the Nez Perces' assurances of peaceful intent and not wanting to risk fighting in their own back yard, urged caution on Rawn and decided to return to their homes. Except for minor thievery and destruction here and there, the Indians spent eight days peacefully moving up the valley, trading with the settlers as they went. Only when they exited Flathead territory did they vent their anger by breaking into M. M. Lockwood's house, the last one in the valley, and demolishing everything in it but the stove. Rawn was criticized in some quarters for his timidity, and his barricade was derisively named "Fort Fizzle". But the wisdom of his caution was soon to become evident to even the most virulent of Indian haters.[11]

General Gibbon, who arrived at Fort Missoula on August 4, immediately sought out Father Ravalli at St. Mary's Mission to pay his respects, but also to gather intelligence on the hostile Nez Perces. Ravalli, who was bedridden, told Gibbon, "They are splendid shots, well-armed, have plenty of ammunition and have at least two-hundred and sixty warriors." He warned, "You must not attack them, you have not enough [men]."[12]

But Gibbon would not be deterred. He was a man in a hurry, a man with a mission. He would not "pull a Rawn". He not only had his orders, he was determined to avenge the set-backs Howard had suffered in Idaho. He was going to make sure that the Army's prestige and honor, sullied at the Little Big Horn, would be restored and the Nez Perces be punished for the death and destruction they had caused in Idaho. He had little use for the Bitter Root settlers and accepted volunteers with great reluctance.

Chief Looking Glass, on the other hand, convinced that he could proceed peacefully through the Big Hole and Beaverhead basins on his way to the Yellowstone River, was lulled into complacency by the peaceful trek up the Bitter Root. He resisted all efforts by the other chiefs, including Joseph, to hurry the caravan along, considering it more of a traditional buffalo hunt than a flight for the very life of his people. When they finally made camp on the North Fork of the Big Hole River at the junction of Trail and Ruby creeks on August 7, he insisted that it was unnecessary to post guards or lookouts and he vetoed an effort by younger braves to send scouts back along the trail to see if they were being followed. Not even a picket was set out to stand watch over the horses. By the night of August 8, Gibbon had caught up with them as they rested peacefully at their campsite. With seventeen officers and one hundred and forty six enlisted men, he decided to launch a surprise attack at dawn.[13]

It is hard to know what Gibbon hoped to accomplish by this attack and the record doesn't shed any light on it. Had he reflected on alternatives or did he just race ahead mindlessly? It is clear only that his mind was set on a fight. He said, "Take no prisoners." That means that even if the Indians had surrendered they would have all been slaughtered. Did he really intend to wipe out the entire caravan, men, women and children with his attack?

With the first light of dawn Gibbon moved his men forward. When they were within about two hundred yards of the line of Indian tepees, a single Indian brave on his way to check on the horses walked into them without realizing it. He was shot down instantly and the attack was on, with the troopers and volunteers running forward shouting loudly and shooting low into the Indian tepees in order to inflict maximum casualties. There was wild panic among the Indians awakened from their sleep and running from their tepees in various stages of undress. With screams and mad scrambles they scattered, looking for their weapons and running for the banks of the creek for cover or back into the trees. The fighting was fierce and at close quarters, the soldiers using bayonets and rifle butts as well as bullets and the Indians fighting back with axes, clubs and rifles.

The battle raged for four or five hours with the Indians gradually getting the upper hand. Gibbon, who had been wounded in the thigh, then withdrew his men to a low hill south of the action. While they were digging in they were suddenly surrounded by whooping Indians firing into their position from all sides. Shouts of "It's another Custer Massacre," were heard. The battle had completely reversed itself.

The Indians kept Gibbon's position under siege for the next twelve hours. They could have overrun it, wiping out the entire command, but they had other priorities. They had to get their caravan together, take care of their dead and wounded, many of whom were women and children, and they didn't want to sustain the casualties that it would take to overrun Gibbon's position. Gibbons, for his part, managed to dispatch three messengers during the night, one to General Howard and two to Deer Lodge via French Gulch. One of them, W. H. "Billy" Edwards, arrived in Deer Lodge two days later. [14]

Finally, at dawn the following morning messengers from General Howard arrived at the scene. The Nez Perces correctly interpreting this as a sign that troop reinforcements were on the way broke off the siege and rode off to catch up with the caravan which had begun its sad and painful departure hours before.

The battle had lasted exactly 24 hours. Gibbon had suffered heavy losses, thirty-one soldiers dead and forty wounded, with seven of his seventeen officers on the casualty list, two dead, five wounded. The Indians were also badly hurt. Between sixty and ninety Nez Perces had lost their lives, most of them women and children hit in the initial attack. But they had also lost from twelve to twenty of their best warriors.

It wasn't until the afternoon of August 11, that General Howard's main force arrived at the battle ground with two doctors to attend the wounded. The next day W. A. Clark with thirty-five volunteers, two doctors and four wagons arrived from Butte.

For those just arriving, the battlefield presented a ghastly spectacle. A soldier, Frank Parker, wrote, "I went over the field and the sight was the most horrible and sickening I ever beheld. The banks of the creek...was literally lined with festering, half-putrid corpses..." Major Edwin C. Mason, one of Howard's officers wrote, "...—it was a dreadful sight—dead men, women and children. More squaws were killed than men. I have never been in a fight where women were killed and I hope never to be."[15] Mason may have been spared such an experience, but what happened in the Big Hole was certainly not unique in the long history of Indians resisting the white invaders.

The volunteers from Deer Lodge Valley, who had been instructed to guard the Mill Creek Canyon pass from the Big Hole, after waiting a week on upper Mill Creek without hearing word of the Indian advance, had returned to their ranches in the valley just as Gibbon's messenger Edwards arrived in Deer Lodge. They were told to reassemble and organize a train of ambulance wagons and join the doctors and nurses who were on their way from Helena and Deer Lodge. The ranch houses along Mill Creek were to be used as rest stops and way stations between the battle ground and the newly constructed St. Joseph's hospital in Deer Lodge.

This party with twenty wagons and three doctors arrived at the battle scene on August 13, and immediately began evacuating the wounded. For the next week the Evans house became an emergency hospital where wounds were dressed and wounded cared for until they could be moved to Deer Lodge. Legend has it that Marcus Daly participated in this evacuation of the wounded and it was at that time that he met and became friends with Morgan Evans.

Howard with his entire command of two hundred and twenty-seven regulars and a large group of packers, guides and volunteers departed that same morning to renew his pursuit of the Indians. Those Bitter Root volunteers who had survived the battle (some had run off during the fighting) had by then already returned to their homes, but the Butte volunteers under "Major" Clark, and some from Deer Lodge under "Captain" Stuart rode off with Howard. However, according to Thomas A. Southerland, a correspondent from the San Francisco *Chronicle,* traveling with Howard's forces, "Sixty volunteers from Deer Lodge, who joined us last night, left this morning because they were unwilling to submit themselves to military discipline." Actually, this included both the Deer Lodge and the Butte volunteers. "Captain" Stuart explained to Secretary Mills at Deer Lodge a few days later that W.A. Clark had rankled General Howard by trying to get him to divert cavalry to support the volunteers in a hasty firefight with the retreating Nez Perces.[16]

The danger of an Indian attack on settlers in Deer Lodge Valley had passed, but the possibility of continued fighting further east remained. For the Indians, the danger was complete annihilation; while settlers along the way feared Indian retribution. The Nez Perces caravan moved slowly after the battle, taking care of their wounded as they went. They traveled along the western edge of the Big Hole valley and crossed the mountains into the adjoining small valley of Horse Prairie Creek, near Bannack, where they attacked some farmers haying, killing four of them, ransacked the farm house and stole their livestock. Further south at Birch Creek, they attacked a wagon train en route from the railhead at Corrine, Utah carrying supplies to Salmon City. There were eight wagons, thirty mules, three drivers, four passengers and two dogs. Among the supplies was between 3,000 and 10,000 rounds of ammunition for the Salmon City volunteers and ten barrels of whiskey. It was later reported that the Indians, drunk on the whiskey, killed the three drivers and two of the passengers, burned the wagons and supplies and stole the mules. The two other passengers, both Chinese, were spared but were left to make their own way to the nearest settlement.[17]

By August 19, Howard and his troops arrived at Camas Prairie close behind the Nez Perces' caravan. This time it was the Indians turn to make a surprise attack. But still wanting to keep their losses to a minimum they decided to stage a raid on the General's horses to slow down his pursuit, rather than on his camp. They got away with two hundred mules and a number of horses. The ensuing battle left one soldier dead and eight wounded. The Indians suffered only a few slight wounds.

After evading the pursuing army as they crossed through Yellowstone Park, the Nez Perces were disappointed to learn that the Crow nation did not want to give them refuge and that Sitting Bull was in Canada, so they headed north for Canada. By this time the pursuit had become a national drama, with the Army and General Howard being held up to ridicule. General Sherman, who had arrived at Fort Shaw, assigned Colonel Samuel D. Sturgis with six companies of Custer's old Seventh Cavalry, four hundred men strong, to Howard's command.

Sturgis finally caught up with the Indian caravan near Canyon Creek just west of Billings on September 13. The Indians had just looted and burned a stage station and made off with the stage coach and horses. In a skirmish that lasted most of the day, Sturgis lost three men killed and eleven wounded. The Indians once again had been able to out-maneuver and out-shoot the cavalry vanishing into the night with only three wounded braves. But the Indians were being worn down. It had been three months of fighting and running, of losing women, children and older members who couldn't keep up, as well as fighting men and horses. The strain was beginning to tell and many were losing heart.

Since they had easily out-distanced Howard and Sturgis in their flight northward, they made the mistake of stopping to rest forty miles short of the Canadian border, on Snake Creek near the Milk River on the northern flank of the Bear Paw mountains, confident that they could now make it.

But the Indians were at a definite disadvantage. The U.S. Army, with its use of the telegraph, could call up reinforcements from afar. Unknown to the Nez Perces, another Army unit had joined the fray under Colonel Nelson Miles with three companies of the Second Cavalry, three of the Seventh Cavalry and five of the Fifth Infantry, four of which were mounted, three hundred and eighty three men in all. Miles had left Fort Keogh on the Tongue River on September 13.

On September 30, Mile's Cheyenne scouts finally located the Nez Perces camp. He ordered an immediate attack before they could get away. Although the Army suffered heavy casualties in the initial attack, it was able to surround the Indian camp. Unable to overrun the Nez Perces positions, Miles laid a siege that lasted five days. By the time Chief Joseph surrendered on October 5, and made his famous statement, "I will fight no more, forever", General Howard with his entire command had arrived to reinforce Miles.

This ended the Nez Perces war. They had battled some 2000 Army regulars and volunteers in a 1700-mile retreat over a period of three-and-a-half months. They had at least one hundred and twenty of their people killed including women and children and they had slain approximately one hundred and eighty whites and wounded one hundred and fifty. Miles took four hundred and eighteen battlefield prisoners: eighty seven men, one hundred eighty four women and one hundred forty seven children. Approximately two hundred and thirty-three others slipped through Army lines and escaped. Suffering from cold, hunger and fatigue some died on the plains from exposure, others were killed by hostile Indians, but most of them made it into Canada and Sitting Bull's protection.[18]

Having spent a winter in Canada, a small group of embittered Nez Perces survivors, who had decided to return to their traditional lands, made its way south, down the Blackfoot River across the Deer Lodge and up Rock Creek in the vicinity of Philipsburg. They killed two miners and took several hundred dollars in gold dust in Bear Gulch just before crossing the Deer Lodge River. Before arriving at Rock Creek, they came upon the cabin of two prospectors. The chief of the band questioned them as to whether they were soldiers or had participated in the Big Hole battle. After receiving fervent denials, they departed appropriating meat and other provisions they found in a shed against the cabin.

Two days later, July 11, 1878, they killed John Hayes, another prospector, outside his cabin in McKay Gulch on upper Rock Creek about twenty miles west of present day Anaconda. The next morning about day-break further up the gulch they burst into the cabin of three other prospectors, Amos Elliott, William Jory and J.J. Jones. Elliott and Jory were killed and Jones made his escape with a bullet wound in the shoulder. He finally shook off his Indian pursuers and made it into Philipsburg late that night, thoroughly exhausted and weak from the loss of blood.

These murders caused panic in the surrounding countryside and great excitement in Philipsburg. Ranchers packed up their families and crowded into town for protection. Miners at nearby Georgetown and Southern Cross began round-the-clock armed surveillance of their properties. A posse was organized under captain John

McLean, who recovered the bodies but found no trace of the Indians who had moved on. Jones recovered fully from his wound. He became a local legend as the man who outran the Nez Perces and was known thereafter as "Nez Perce" Jones. The Philipsburg mortuary record lists Amos Elliott—killed by Indians, McKay Gulch, 1878; John Hays—killed by Indians, McKay Gulch, 1878; William Jory—killed by Indians, McKay Gulch, 1878. These were the last known casualties of the Nez Perces War.[19]

However, Custer's defeat and the long campaign against the Nez Perces created new pressures to increase the army's presence throughout Montana Territory. Prior to the Little Big Horn battle, there were about six hundred soldiers in Montana. At the time of Joseph's surrender, there were three thousand and pressure from the public was strong to increase that number. As a result, eight additional forts were built between 1877 and 1892.[20]

7. Clark, Daly and *The Anaconda Standard*

Before the gold stampedes in the Northwest in 1861 and 1862, Washington territory stretched from Puget Sound to the continental divide. Everything between the divide and Minnesota was Dakota Territory. Idaho Territory, created in March 1863, lopped off Washington Territory at the bend in the Columbia River near Walla Walla and extended through present day Montana, Wyoming and part of Nebraska.

When the gold discoveries at Bannack, Virginia City and Helena caused a sudden increase in population, there was a clamor for the creation of a separate territory with an administrative capital close to the new mining activity. Former Ohio Congressman Sidney Edgarton, assigned to serve as Chief Justice of Idaho in its eastern district at Bannack and his nephew, Col. Wilbur F. Sanders, both recent arrivals, became early promoters of the idea. With the assistance of Sam Hauser, Francis M. Thompson and Nathaniel P. Langford, they raised $2,500 in gold samples and additional cash to send Edgarton to Washington to petition for territorial status. Edgarton was to use the gold to impress the U.S. Congress with the wealth of the area.[1]

Upon his arrival in Washington in January 1864, Edgarton was pleased to find that another Ohio Congressman, James M. Ashley, had already introduced legislation to create the new territory. Montana was the name that Ashley had chosen for it and he had designated its western border, not on the continental divide as might have been expected, but along the mountains to the west of the Bitter Root Valley. With only minor opposition, both the Senate and the House passed the bill by May 20 and six days later President Lincoln signed it into law.[2]

It is worth noting here that since the South had seceded, Congress was dominated by the Republicans, who were anxious to increase their hold on the West and eventually increase their numbers in Congress through additional representation from new territories and states. Those sponsoring territorial status for Montana, Ashley,

Edgarton and Sanders were all hard-line Radical Republicans who viewed Democrats as secessionists and traitors. Nonetheless, a large majority of the approximately fifteen thousand inhabitants of the new territory were Democrats, many having migrated from the South, often by way of Missouri and the border states. Sanders observed that Montana was "in the hands of refugees from Price's Army, Missouri having the honor to mother over half of our voting population at least..."[3]

Others, such as the immigrant Irish, viewed the Republicans as anti-immigrant and anti-Catholic therefore identified with the northern Democrats. This division was to fuel a running battle between the executive and legislative branches in the new territory throughout its entire 25 year history. While a succession of Republican Presidents, appointed territorial Governors who were Republican, the territorial legislatures were heavily Democratic and the territorial delegates to Congress were often Democrats too, though some of the Governors strove mightily to change the Territory's political complexion.

Sidney Edgarton was appointed the first Governor of the Territory in July, 1864. Tiring early of his fights with a Democratic legislature and frustrated by lack of timely funding from Washington, he left in September 1865 when Thomas Francis Meagher arrived in Bannack to take up his duties as Territorial Secretary. Meagher, an Irish immigrant who had recruited and led an Irish brigade for the Union Army, was a Unionist Democrat appointed by Andrew Johnson after Lincoln's death. Edgarton never returned to Montana in an official capacity and Meagher became acting Governor until Green Clay Smith was appointed Governor in 1866. Twenty-three years later, in 1889, Montana became a state as part of an omnibus bill, in which Washington, North and South Dakota were also admitted to the union as states. Idaho, Wyoming and Utah became states the following year.

At the time Montana achieved statehood its two largest and most important cities were Butte and Helena. Helena boasted of having produced fifty millionaires and was said to be the richest city per capita in the United States. While the new state was sixty percent rural, the mining and mineral wealth in its southwestern corner were responsible for it having become first a territory and then a state, and most of the population was concentrated in and near those cities.

Public life and the political agenda were directed from Butte and Helena and dominated by the mining interests. In Helena, it was Charles Broadwater and Samuel Hauser who called the tune. In Butte it was W.A. Clark and Marcus Daly.[4]

These were the big four of Montana politics and economic life. Although all four considered themselves Democrats and were prominent in party affairs, their economic interests dictated their day-to-day political leanings. As might be expected, the mining interests, headed by W.A. Clark and supported by Marcus Daly, completely dominated the state constitutional convention of 1889 and were able to fashion the new state constitution to their own liking, providing special tax advantages for mineral production.[5]

But these two men did not often work in harness or in harmony. They had had a serious falling out in 1888 when Clark, a Democrat, lost the race for Territorial delegate to the U.S. Congress to Republican, Thomas H. Carter, a political unknown. It has never been established with certainty exactly what happened, but Daly didn't deny having had a hand in it. Whatever the case, it was probably not an easy decision for Daly, who did business with Clark regularly, and whose brother, J. Ross Clark, had married Daly's wife's sister.

The relationship between the two men had never been a warm one. There are indications that Clark, perhaps because he considered Daly's successful mining operations a threat to his own, had taken an early dislike to Daly. Clark, who had pretensions to status and refinement, considered Daly below him intellectually and socially, and soon after they had met, had written letters to both the Walker brothers and Haggin trying to discredit Daly. Daly, who was aware of these letters and probably of Clark's feelings toward him carried on a correct but less than cordial relationship with him. There are a number of theories as to what inspired Daly to assist in the anti-Clark effort. One, which has been repeated over the years, and which one of Daly's daughters believed, was that Clark had tried to tie up all the water rights on Warm Springs Creek when Daly began buying up land for his new smelter and town. While there is nothing in county records to sustain that Clark was involved in such an effort, he did have interests on the creek at the Blue Eyed Nellie silver mine before Daly had arrived on the scene. This was just five miles west of the site Daly chose for his first smelter. It could very well be that

Clark, operating through mine owner Frank Brown, or third parties, tried to choke off Daly's smelter project by denying him the water which was essential to its operation and which had attracted Daly to the site in first place.[6]

However, another theory, advanced by historian K. Ross Toole, has gained currency in more recent writing on the subject. It contends that Daly's decision was the result of legal action brought by the U.S. Department of Interior against Daly and his Missoula partners in the Montana Improvement Company.

From 1885 to 1889, the Cleveland administration, through the Secretary of Interior, General Lucius Lamar, and his Land Office Commissioner, William Andrew Jackson Sparks, conducted extensive investigations into fraudulent practices of cattle ranchers, railroads and timber companies on public lands and squatters on Indian reservations and pursued a vigorous campaign against predatory interests that were despoiling lands and forests in the West.[7]

A particular target of Sparks' investigation was the Montana Improvement Company, jointly owned by Daly, the Northern Pacific Railroad and lumbermen Hammond, Bonner and Eddy of Missoula. Sparks instituted both criminal and civil suits against the company for fraud and trespass on government lands and the wholesale destruction of government-owned forests. Although the company had been formed to cut timber on Northern Pacific lands obtained in the railroad's original land grant, it preferred to hold on to the good timber on railroad lands waiting for higher prices in the future and cut trees on government land instead. The Company had no interest in obtaining a license from the government for this activity since even if it had been granted, it would have added at least $500,000 a year to the Anaconda Company's timber bill.[8]

As a result of its campaign, the Cleveland administration reclaimed some eighty million acres of public lands, but the uproar from railroad, mining and cattle interests was so loud that Sparks was forced to resign in November 1887. Nevertheless, many of the suits that he had initiated stayed on the dockets, including those against the Montana Improvement Company.

In the spring of 1888, Clark became a candidate on the Democratic ticket for territorial delegate to the U.S. Congress. Because of the concentration of southerners and Irish in western Montana at the time, it was overwhelmingly pro-Democrat, and Clark assumed he would win an easy victory. While he may not have expected Daly to actively support his first political bid, he certainly didn't anticipate his opposition.

But the federal government's suit against the Montana Improvement Company worried Daly and his associates. They were looking for high-level political remedies. Having concluded that the Republicans would take the White House in the 1888 elections, they believed it was important for them to have Republican friends in the right places. So, Eddy, Hammond and Bonner, all avowed Democrats, together with officials of the Northern Pacific railroad urged Daly to join them in backing the Republican candidate. When Daly agreed, word went out to chiefs of train crews, mine foremen, smelter superintendents and timber bosses that they were expected to round up votes for Carter. This was in the days before the secret ballot when bosses could, and did, examine ballots before they were cast. Thousands of employees of the Anaconda Company, the Montana Improvement Company and the Northern Pacific railroad turned in ballots on which the name of Carter had been pasted over the name of Clark. Clark was not only defeated but stunned by the size of the Republican majorities recorded in normally Democratic counties, including his own Silver Bow county.[9]

David Emmons in his 1989 book, *The Butte Irish*, questions this thesis, indicating that the reasons for Clark's defeat went well beyond any effort by Daly to deprive him of elected office. He points out that the 1888 election of Republican, Benjamin Harrison, as President, was in no way predictable. In fact, Grover Cleveland actually won the popular vote, but was defeated in the electoral college. It was widely commented afterwards that the outcome of the election turned on a major diplomatic gaff by the British Ambassador in Washington. He is reported to have publicly stated that President Grover Cleveland would probably be more friendly toward Britain than would Benjamin Harrison. This is said to have turned the Irish vote against Cleveland and decided the election. In Montana, another factor was the Democratic party's unpopular position on the tariff. Clark, supporting his party's position, campaigned against the tariff, thereby losing votes. In addition, Emmons reminds us that Thomas Carter, even though he was a Republican, was an Irish Catholic. Clark, on the other hand, was a Scotch-Irish Presbyterian, an Orangeman, a Grand Master Mason, and many suspected anti-Catholic and anti-Irish independence. These fears were stimulated, if not confirmed, by the Clark campaign which carelessly slighted Irish Catholics in Butte by scheduling a major campaign event, a beef barbecue, on a Friday, a meatless day for Catholics, and inviting to this event members of Sons of America, a virulently anti-Catholic organization. Clark also made disparaging remarks in a speech at Missoula against Patrick Ford, editor of the *Irish World*, a strongly pro-Irish independence national journal. Thus, Emmons asserts, there were plenty of reasons for large blocs of voters in both Butte and Anaconda to neither like nor trust Clark and to vote against him without being influenced or coerced by Daly. However that may be, Clark had no doubt who was to blame for his defeat and from that time forward openly vilified Marcus Daly.[10]

On Daly's side, the feud with Clark went beyond the two of them. It is said that Daly's partner, Haggin, had an abiding hatred for Clark, who had once called him a "nigger", alluding to his Turkish mother. H. Minar Shoebotham in his *Life of Marcus Daly* reports that a year or two before Daly's death when it was suggested that he "call off this damn fool feud with Clark", he replied, "Ask Mr. Haggin about it some time." Haggin, who outlived Daly, is reported to have said, "As long as I have any interest in the State of Montana, W.A. Clark will never be a senator."[11]

As for Thomas Carter, he lost no time in repaying his benefactors for getting him to Washington. In just over a month after he took his oath of office, the Secretary of the Interior John W. Noble instructed the U.S. District Attorney for Montana to suspend all actions against the Montana Improvement Company and the Northern Pacific railroad. Carter made sure everyone knew of his handiwork by extolling Noble a few days later in an interview in the Republican, Butte *Intermountain* newspaper.

For Carter, the election was the beginning of a prominent career in the Republican party. He gained a reputation as a staunch pro-business conservative who worked hard for Marcus Daly and Montana mining interests, became a close personal friend of President Benjamin Harrison, was his campaign manager in the 1892 elections and national chairman of the GOP. He eventually served two terms in the U.S. Senate, 1895-1901 and 1905-1911, was a strong advocate of tariff protection, a major force in creating Glacier National Park and establishing the postal saving system, as well as being opposed to the creation of national forests.[12]

For Clark it was a public humiliation which he never got over and for which he never forgave Marcus Daly. Even though in 1889, he worked with Daly and the other mining interests to get favorable tax treatment for mineral exploitation written into the state constitution, he simultaneously mounted a noisy campaign through his Butte newspaper, the *Miner*, to vilify Daly, Haggin and the Anaconda Mining Company.

Daly's reaction was predictable. He would start his own newspaper and overshadow Clark's "rag". Since he didn't believe in doing anything by halves, he wanted only the biggest and the best. He decided that his newspaper would be published in his new town, Anaconda, because Clark controlled the Associated Press franchise for Butte.

As luck would have it, at the time Daly made his decision, Dr. John H. Durston, a PhD. graduate from the University of Heidelberg, a professor at Syracuse University and ex-editor of the *Syracuse Standard*, was in Anaconda visiting his friend Frank Leonard who published *The Anaconda Review*. After reading an editorial that Durston wrote in the *Review* that impressed him, Daly proposed to Durston that he establish and run his newspaper, which would be the "best in the West". Durston was dubious. He didn't believe the town of Anaconda, with a population of only 5,000 people, could support a large circulation newspaper, even though Daly offered to cover its losses. Daly insisted that the paper would not be just another small town newspaper, but a regional paper that would gain respect and be sold all over western Montana and that it would compete directly with the Butte papers, the *Miner* and the *Intermountain.*

The story goes that even after Durston studied the economic and business situations of Anaconda and Butte and was favorably impressed by the growth potential of western Montana, he was still hesitant until he read in the Butte paper that Daly had just paid four thousand dollars for a colt from a famous Kentucky stable. When he met with Daly, he accepted his offer with this commentary, "If you can sink $4,000 into a colt that may or may not become a race horse, I guess you can put $40,000 into an outfit that may or may not become a newspaper."[13]

The *Anaconda Standard* published its first edition on September 4, 1889. Over the next few years, Daly spent ten times more than the $40,000 Durston had mentioned on his new *Anaconda Standard*. He instructed Durston to hire only top journalists and buy the best and most modern equipment. Most small-town dailies had hand-operated presses and hand set their type, but Durston, on Daly's insistence, ordered automatic typesetting machines, the latest Mergenthaler linotype. It was said that the *Standard* came to have more linotype machines in operation than any of the dailies in New York city. These machines were more than just an idle whim of Daly's. For the *Standard* to succeed, it was absolutely essential that the new paper dominate its regional market. Daly had to be able to publish with speed and in quantity so that his papers could arrive at distant locations in time to match or beat the competition.

The first issue of the *Standard* declared its politics and purposes in these words: "Here goes a daily newspaper. It is the vigorous child of a wide awake town...It has been christened the *Anaconda Standard*. It will greet the public every morning...The newspaper declares itself a Democrat...It appears that a very large majority of the men who built the state are Democrats...Its home is in a city of mechanics and laborers..." In a story on another page it announced that the *Standard* occupied the best business block in town, three stories high, sharing quarters with the power company, the water company and the Rocky Mountain Telegraph Company.[14]

In the spirit of the times, it took its political commitment as Democrat seriously. Even though Daly's primary purpose in founding the paper was to create a rival for Clark's *Miner*, that paper too was Democrat. Consequently, *Standard* editorials, more often than not, attacked its Republican rivals, the Butte *Intermountain* and the weekly *Anaconda Review*, rather than the *Miner*. More importantly, *The Standard*, like its rivals, carried partisanship to the extreme of not reporting anything favorable about opposition party politicians. At election time it did not even report the names of the candidates of opposition parties; only names from the Democratic party slate appeared in its news columns and editorials. And only reluctantly and with minimal space did it report the names of election winners who didn't happen to be Democrats.

Each issue of the *Standard* sold for five cents, or $1.00 a month subscription. This was expensive, compared with daily papers in New York City, which sold for one or two cents a copy. Utilizing the full services of the Associated Press, daily editions had eight pages of local, national and international news in approximately equal proportions and used the standard format of the time, six closely packed columns on each page with headlines above the story in each column. And, as is the case in some European newspapers to this day, classified advertising was prominently presented down the left hand column on page one. For the first six months, six issues a week were published, skipping Mondays, as was the custom among other papers in the West at the time. However, by April of 1890, it had become so popular, that it began a Monday edition and proudly added, "Published Every Day of the Year" to its masthead. At this same time, *The Standard* expanded from eight pages to twelve for most editions, and sixteen pages on Sunday.

The inside pages carried extensive advertising, mostly from Butte and Anaconda, giving us some idea of the flavor of the two cities. The largest advertisers were mercantile companies, which carried both dry goods and groceries, like D.J. Hennessy, which had stores in Butte, Anaconda and Missoula; Estes and Connell and McCallum & Cloutier which had large stores in Anaconda.

The following advertisements all taken from a single early issue (Thursday, October 17, 1889) convey the sense of growth and bustle the town was experiencing:

The Anaconda Natatorium advertised its "plunge bath three to seven feet deep on West Second Street one block from the Montana Hotel, George Savage, manager". John Petritz had a large ad as a wholesale dealer in Fine Kentucky Whiskey, Wines and Liquor and J.M. Martin announced his "Specialty in Fine Cigars, Domestic, Imported and Key West". W. Stephens advertised as a Physician & Surgeon; F.P. Chrisman "D.D.S. Teeth Extracted Without Pain"; J.D. Fitzgerald, "Public Magistrate, Justice of the Peace and Notary Public"; J.C. Keppler, "Watchmaker and Jeweler"; E. C. Freyschlag & Co. "Leading Shoe Dealer"; Beilenberg & Co. "Beef, Mutton, & Pork"; Sam Pramenko "Dealer in Fresh Game, Oysters, Fish, Fresh Eggs and Fine Liquors". And M.S. Aschheim announced the grand opening of his "Mammoth Dry Goods Emporium". A.T. Playter, who had a drug store on First Street near Main, advertised daily in the *Standard* and also provided the thermometer readings reported each day in the paper.[15]

Since it was meant to be a regional paper, *The Standard* covered events in Butte and surrounding communities in detail, but Anaconda was not neglected. We are informed on page four of the first issue that "Lumber has been sawed into sidewalks during the past two months by the hundreds of thousands of feet. The city is now well supplied with excellent highways for pedestrian traffic", and, further on, that the contractors building the new sewer have almost finished digging trenches for the pipe. It also reported that F.G. Brown's Blue Eyed Nellie mine was producing forty ton of ore a day, that it was 500 feet down, and that it had fifteen men on the payroll.

In another section we read that the "City council discussed whether permits should be granted to allow cattle to graze within the city, and that "all property owners are required to put in sidewalks." It also announced that "Deputy Clerk Mahan will be in Carroll from 10 in the morning till 10 in the evening to prepare naturalization papers for all who apply." [16]

Under the able management and editing of Durston, *The Anaconda Standard* became, as Daly had wanted, the leading newspaper in Montana and the Northwest with a circulation of 20,000. It quickly became the largest selling newspaper in Butte and it proudly announced that it had a larger circulation in Philipsburg, Granite, Missoula and the Bitter Root than all of its rivals combined. Only the *Oregonian* of Portland had a larger circulation in the Northwest at the time. *The Standard* worked hard at keeping its readership and increasing its circulation. It made special arrangements to get its daily editions to the Flathead, where there were still no railroads.[17]

During the suspension of service on the Montana Union rail road between Butte and Anaconda for five months in 1891, *The Standard* contracted a special locomotive to carry papers every morning to Butte, and also to make connection with the UP trains for morning delivery to cities in Idaho and northern Utah. As an alternative, it also experimented with relay teams of horses, pulling wagons from Anaconda to Butte loaded with the early edition. When full service on the railroad was restored, *The Standard* began to print two early editions. The first was loaded on the Northern Pacific train every morning at 2:15 a.m. at Stuart destined to reach Deer Lodge at 3:30 a.m., Garrison at 4:00 a.m., Missoula at 6:40 a.m. and Ravalli, for delivery in the Flathead, at 8:22 a.m. The second edition was sent to Butte at 4:00 a.m. for local circulation and to meet the connections for southbound trains.[18]

Within a few short years, *The Anaconda Standard* had all the characteristics of a large metropolitan daily offering as much national and international news as local coverage, and was soon recognized throughout the United States as an outstanding newspaper. By association, it added class and distinction to the city of Anaconda.

Durston brought in two of his associates from the *Syracuse Standard*, Charles Eggleston and Warren H. Wallsworth, as associate editors, and Charles Knox to run the engraving department. When colored comics made their debut in New York city with "The Yellow Kid", Daly wanted the same for his paper. He hired Thorndyke, Trowbridge and Loomis, three of the best known and highest paid newspaper artists in the U.S. John and Paul Terry were also artists and cartoonist on the paper. He ordered color decks and photo-engraving equipment and published a four-page colored comic section on Sundays.[19]

As long as Daly lived, he lavished as much money and attention on the *Anaconda Standard* as he did on any of his thoroughbred horses and got as much personal joy out of seeing it win awards and gain renown. In some circles, he was better known for his newspaper than for his other activities. He must have been amused to see himself, with his limited education, referred to by other newspapers as a publisher, rather than a miner. When he died the New York *Telegram* didn't even mention his industrial achievements but referred to his stables and continued: "He established one of the best newspapers in the country, the *Anaconda Standard*, in a mining town remote from dense population."[20]

He eventually spent $5 million on his paper. The words of praise and admiration it received for the special 56 page Christmas edition of 1898 and the 1899 special edition on the return of Montana soldiers from the Philippines at the end of the Spanish American War gave him great pleasure. But Daly never forgot his primary reason for founding the *Standard*. He would use it in his fight with Clark, who had positioned himself to become U.S. Senator from Montana and who was pushing for Helena to be named the State capital. Daly would oppose Clark's bid for the Senate and make sure that Anaconda would become the capital of the State. The *Anaconda Standard* would help him get his way.[21]

From the day Durston accepted Daly's offer to run the paper, he had worked overtime to begin publishing soon enough to influence the special election of October 1, 1889. This election was to fill all the offices of the new state government, scheduled to come into existence in November. With the first issue appearing only on September 4, there was precious little time.

National attention was focused on these special elections in all four of the "omnibus states" because control of the U.S. Senate was at stake. The national committees of both the Democratic and Republican parties poured money into local elections, too, since at this time, the state legislators still determined who sat in the U.S. Senate. In addition, the big four of Montana politics invested heavily seeking control of the state government.

A strong Republican trend nationwide came within a hair of making it the dominant party in the new state and produced a legislature that was evenly divided. In the end, with a deadlocked legislature, each party chose two Senators and sent them to Washington. The Democrats named W.A. Clark and Martin Maginnis and the Republicans, Wilbur F. Sanders and T.C. Power. The U.S. Senate, which had a narrow Republican majority, naturally seated the Montana Republicans. For Clark, sitting in the Senate gallery watching the vote, it was another bitter blow.[22]

While Daly and his new *Anaconda Standard* had little or no influence over this outcome, he must have offered Durston a cigar in quiet satisfaction.

8. Winning

From the very beginning, one of the defining characteristics of Daly's new town was an intense love of sports, games, contests and competitions of every sort. Daly's ambition was for his town to be the biggest, best and first in everything. Among the citizenry this translated itself into a fierce pride and determination to make its mark and to excel at every endeavor it engaged in. Of course, the love of games was general across the broad expanse of America and was not unique to Anaconda. What did distinguish the town, however, was the fierceness of its determination to win and to be seen as a winner. This was a characteristic that it never lost, even in the worst of times.

In those days Butte as a city was a series of mines and neighborhoods around the mines with separate identities such as Butte Hill, Walkerville, Dublin Gulch, Centerville, Meaderville, and McQueen to name a few. Sports teams, marching bands and booster clubs were organized by the various mines, neighborhoods and ethnic groups. Competitions and tournaments were arranged among them on an ad-hoc basis from season to season. Although Anaconda was twenty-five miles away, it automatically fit into this existing pattern and its teams became part of the regular competition. But Butte also had an identity as a city and fielded strong teams made up of the best from all its neighborhoods. Undaunted Anaconda took on Butte teams of whatever origin and never asked for quarter. Whatever the sport or game organized in Butte, Anaconda would be there as a challenger.

There's no doubt that much of this swagger came from Marcus Daly himself and in the early days was associated with horse racing. Daly's love of the sport has been widely noted and commented on. It was an enthusiasm that he shared with his partner from Kentucky, James Ben Ali Haggin, and his good friend Morgan Evans, who raised Hambletonian trotting horses on his ranch in the valley. It was only natural then that if Butte had a race track, Daly would share his enthusiasm with the inhabitants of his town by building one for them. Thus, the track, built west of town in 1888, became an active part of the city's life,

In its early years, the track had a thirty day season, with trotting races dominating the program. This was an important addition for the town not only because it provided a prestigious recreational facility, but also because it put Anaconda on the racing map. Conversely, Anacondans followed developments on other race tracks around the country, placing bets on horses and following the races through telegraph despatches posted regularly on the chalk board at the Turf Exchange, a bar and cafe on Main Street.

Helena and Great Falls also had one week racing programs during the summer, which were scheduled to extend the total western Montana racing season, and later Missoula and Glendive had similarly short seasons, but the races at Anaconda and Butte were regularly listed nationwide in the racing news and were considered the best in the West. For the next twenty years, there wasn't a town its size in the country that had better racing programs, faster horses or more colorful crowds. Some of the fastest horses in the country raced there.[1]

Daly, of course, wasn't satisfied with bringing big-time racing to his copper domain, he had grander plans. He was convinced that horses bred and reared at the high mountain altitudes would develop stronger hearts, more stamina and greater staying power than those raised at sea level. Several years before the Anaconda track opened, he had bought a small ranch in the Bitter Root Valley, while scouting out additional timberlands with his partner, A.B. Hammond. This ranch became the base from which Daly began efforts to prove his thesis. In the process, he sponsored yet another enterprise that brought new recognition and fame to southwestern Montana.

In 1888 he bought additional land in the Bitter Root Valley and established a stock farm there. He spent over $1 million for blooded horses and brought the best racing stock from England, California and Kentucky. He constructed breeding, racing and training stables. There was an exercise track for use in the summer and for the long winters he built an eight mile covered track equipped with its own heating plant, which maintained a constant temperature in even the coldest weather. He gradually extended his holdings in the Bitter Root to include 22,000 acres, and when he built a mansion on the grounds in 1890, it was clear that he intended to spend whatever leisure time he had near his horses.

Whenever a blooded horse won a major race, he'd try to buy it and over time gained the reputation of having as good an eye for horse flesh as for a promising piece of mining property. One of his greatest acquisitions came in 1890, when he bought four yearlings for $10,800 at the annual sale of the Belle Meade Farm near Nashville, Tennessee. One of these, he named Tammany, no doubt in memory of his introduction to the United States and his working days on the Brooklyn docks. Three years later, Tammany would become famous.

The Daly stables' first success on the national scene, though, came from another horse called Montana, a large, clumsy animal with an ugly gait. It wasn't the kind of horse that would impress race goers in a parade around the paddock. When he was sent east to compete in the Suburban Stakes, the odds against him winning were 40 to 1. Montana trailed until the final stretch when his jockey E.H. (Snapper) Garrison gave him the whip and Montana finished first.

Ironically, Daly who was in Butte that day, lost $40,000 on the race. He had been called to inspect a mine that was having some trouble and left his street clothing in the mine change shack while he went below. The shack caught fire and Daly lost his clothes which contained a $1000 ticket for a bet he'd placed on his horse. The joke was on Marcus and it was such an expensive joke that it made the rounds regularly and became part of the Daly legend.

But it would take more than a single win to establish the Bitter Root Stables in racing journals. It would take Tammany, as handsome and gentle as Montana was ugly and mean, to accomplish that. Tammany's debut on an eastern track came at the Eclipse Stakes in New York where he galloped under the wire at odds of sixty to one and made racing fans all over the U.S. sit up and take notice. The Daly stables had been little known till then, but this and later Tammany victories changed all that. As a three-year-old, he won four of five starts, the Lawrence Realization and the Withers at Belmont Park, New York, the Lorrilard Stakes at Monmouth, New Jersey and the Jerome Stakes in New York. Then in September 1893, he was matched against the famous Lamplighter, the fastest four-year-old in the East. Before 15,000 fans at Gutenberg, New Jersey, Tammany roared past Lamplighter in the final stretch and won by four lengths. The residents of Anaconda and Butte went wild. For them, this wasn't just a horse race, but a symbolic victory, the fastest horse in the West had beaten the fastest horse in the East.

4. Marcus Daly's Race Track located just west of the city limits. For about twenty years there wasn't a town of its size in the country that had a better racing program. Some of the country's fastest horses raced there and put Anaconda on the racing map.

After that victory, Daly, in tribute, had a wooden mosaic portrait of Tammany set in the floor of the main barroom of the Montana hotel. It was made of more than one thousand pieces of hardwood of varying shades to capture his likeness. He also built a special stable for him in Hamilton which he called Tammany's castle.

After a lapse of several years Daly's horses were once again in the money on eastern tracks. His trotter, China Silk won the Kentucky Futurity in 1896. Then in August of that year, Daly's colt, Ogden, foaled and raised in the Bitter Root from stout English stock, went east and won the richest racing prize of the year, the great Futurity Stakes at Belmont Park, taking a purse of $57,289. After his racing days were over, Ogden continued to make money for the Daly stables as a stud and became the grandsire of Zev, who won the Kentucky Derby in 1923.

In the following years Daly horses continued to take prizes. Scottish Chieftain won the Belmont Stake in 1897 and Hamburg won the Lawrence Realization in 1898. When Daly died in 1900, his horses were sold, but the lines that he had established went on winning. Hamburg Belle won the Belmont Futurity in 1903. Artful won it in 1904 and Sysonby won the Saratoga Special the same year and went on to win the Lawrence Realization and the Metropolitan Handicap at Belmont in 1905.

Daly didn't live to see perhaps the greatest product of his stables, the great Colin, out of Pastorella, a mare he acquired in 1898. Colin won the Belmont Futurity, the Champagne Stakes and Matron Stakes at Belmont and the Saratoga Special in 1907. Then in 1908, he won the Belmont Stakes and the Withers Stakes. Colin never lost a race.[2]

But horse racing and following Daly's horses weren't the only sporting recreations in the young town. Every type of competition was turned into a spectacle to watch, bet on, get exited about or at least to comment on. As in most frontier towns, fighting of every kind was a popular amusement. Rings or pits were built to watch and bet on dog fights, cock fights and wild animal fights. From time to time bar room brawls turned into more formal matches with self-appointed handlers and referees. Matt Kelly tells of one such fight going on in the yard of the California House. Everyone in the dining room was crowding out to watch when a brawny young Irishman among them, who had gotten out first, reported, "There's nothing to it. Let them alone. It's just two Cousin Jacks."[3]

Regular boxing with professional fighters was always a sure fire guarantee of a full house. Any of various halls in town could be quickly turned into a boxing arena for such occasions. And, of course, for prize fights of national prominence, Ike Quinn's saloon on Main Street and the Turf Exchange posted the round-by-round results, received by telegraph, on a large chalk board for the betting crowd to see.

It was easy for Anacondans to identify with a second generation Irishman out of Colorado's mining district, young Jack Dempsey, and they bet heavily on him in 1891 when he fought the Cousin Jack from New Zealand, Bob Fitzsimmons at New Orleans for a $12,000 purse. There was general disappointment when he lost, but the attention of the betting crowd didn't miss a beat and quickly turned to a series of cock fights between Butte and Anaconda organized at a pit in the city park. And there was general delight when the Anaconda cocks won four out of five fights. Prize fight fans again put up many of their hard-earned dollars in 1892 when Gentleman Jim Corbett put John L. Sullivan away, and lost even more in 1895 when Corbett, favorite of the Irish, was beaten by Fitzsimmons.[4]

Not only horse racing but any kind of racing was on the same level of popularity with fighting and always drew a good crowd. Besides the horse races west of town, there were dog races, or what the Welsh and the Cornish called "courses", held at the Carroll Athletic park. Matt Kelly says that most of the boarding houses and many private houses in Anaconda and Carroll had a half-dozen greyhounds lying around their porches, which they would race. A live rabbit would be turned loose in the park and a pair of hounds would be unleashed after the rabbit, with a judge on horseback to rate the hounds as to which hound made the rabbit turn the most times or made the kill. The rabbit could get away if he found one of the holes in the fence put there for that purpose.[5]

The bicycle, invented in the 1850's, was just becoming a popular American pastime. Bicycle clubs were formed and regular races scheduled. Because there were no steep hills between Anaconda and Deer Lodge, long distance races between the two cities became a regular event. Also, shorter races called scorchers (because of the speeds attained) were held between Anaconda and Gregson. Special bicycle paths were prepared for these races so as to avoid interference from wagons and buggies on the regular road. Thus bikers could streak along without fear of frightening the horses.[6]

The volunteer fire companies would also organize speed trials and races of various kinds to demonstrate their skills in getting to a fire, connecting their hoses and putting the fire out. These competitions were held locally as well as between different cities. One trial consisted of a race of one hundred yards, connecting the hose to the hydrant, then laying one hundred feet of hose. There were also tests of speed and endurance. Johnny Cannavan, a volunteer fireman, held the record for many years for running from the fire station on Oak street to the top of Burnt Hill and back in forty-five minutes. His record was finally beaten by Percy Ingalls, a cop. A surveyors instrument was used to watch that the racers actually went around the flag planted on the top of the hill.[7]

Bowling and swimming were also popular from the time of the founding of the town. Even while tents provided the town's only shelter, makeshift bowling greens materialized to help pass the time, and one of the first saloon buildings had a bowling alley in the back. The mostly single smelter workers formed bowling teams and bowling leagues and were soon challenging Butte and other cities. Because of the sport's popularity, widespread participation and the quality of its teams Anaconda produced a number of outstanding bowlers and became a frequent host for championship tournaments.

While swimming was always more recreational than competitive, it was a summertime favorite of the town's young boys who turned nearby streams, lakes and ponds into improvised swimming venues. And it wasn't long after the city got running water in 1886 that George Savage opened his "plunge bath, three to seven feet deep" on West Second Street to provide year-round indoor bathing to both young and old. On a grander scale, Gregson Hot Springs and hotel were developed in those years offering indoor and outdoor hot pools throughout the year.

Football, both American and Gaelic, and baseball were also played. In the early days most of the ball games were pick up teams for picnics and special events. Organized ball was slower in coming. The first organized football game to be played in Montana was between a Butte team and the Montana College of Deer Lodge in 1893. The Butte team was what today might be called semi-professional made up of local boys and out-of-town itinerant players from college football teams or big city athletic clubs. The newcomers were given regular jobs in the mines or elsewhere to induce them to stay and play football. Butte had ambitions of establishing a team or teams which could challenge the best clubs in the country, which it did in subsequent years. It quickly gained a name in the Northwest by beating such teams as the Omaha YMCA, Spokane, Salt Lake City and Portland as well as Iowa State and the University of Nebraska. From 1893 to 1899, Butte played 36 games, won twenty-three, lost nine and tied three. At the time football was played in both the Spring and the Fall, the forward pass had not been invented and a touchdown counted for four points.[8]

Anaconda followed Butte's lead and organized its first team in 1897. Helena had fielded a team in 1894 and was playing Butte and least twice a year. The Anaconda team played one game with Butte its first year and was beaten 26 to 0. The next year, determined to make a better showing, it put on a recruitment drive and attracted men who had played at the University of California, Stanford, Yale

5. Early day football team, about 1905. Left to right, **back row**: Davis, F. Sharp, R. Lewis, E. Fedell, C. Strokal, S. Nolan. **Middle row**: J. Morris, Sullivan, Dooley, Riordan, Moran, Corcoran, Doran. **Front row**: Marcile, Walsh, Ericson, Corcoran, O'Donnell, Balkovatz. Identified by Bill Keig and Frank Bresnahan.

and Cornell giving ten of them jobs on the smelter. Some of the players had been on West Coast athletic club teams which had been beaten by Butte the preceding seasons and were eager to even the score. This combination of local talent from the previous season and the new recruits was known as the Anaconda All Stars and was immediately recognized as a serious threat to Butte's predominance. The *Montana Standard* in a 1931 article reminiscing about the old time clubs, said that the "Anaconda team of 1898 was one of the best football aggregations that ever stepped onto the gridiron in the northwest or anywhere else."[9]

Anaconda and Butte played four games that year. In the first, on Memorial Day, Butte won 8 to 4. The next game was played in Anaconda on June 13, Miners Day. It drew a crowd of 1,500, 400 of whom had come from Butte on a special train. It ended in a 4 to 4 tie. The following game was played on July 4. Frank Kleptko, Smelter General Manager, contracted the Montana & Boston Band to play at the event. The band had gained national fame two years earlier by playing at the Democratic Party national convention. Anaconda won 9 to 4 and the town was euphoric. At last Dame Anaconda could take her place in the football pantheon.

The last game of the year between the two rivals was played on Thanksgiving day and Anaconda won again 6 to 5. But the referee declared the game null and void because just before the closing whistle with Anaconda in possession of the ball on its twenty yard line someone bounded onto the field, grabbed the ball from the referee's hands and kicked it over the fence.

The following year most of the itinerant stars had departed, but by then there was more than enough local talent to take their place. Many had practiced with the team the year before and had become excellent players. Among them were Jack Livingston, smelterman, guard, 180 pounds; Bob Emmons, library employee, guard, 180 pounds, Bob's brother, Sam Emmons, rancher in the East Valley, half-back and sprinter at 175 pounds. Jack LaFontise, smelter worker from Butte, who'd been acknowledged as the best all-around player on the '98 team, and played equally well on the line or in the backfield; and Jack Sullivan, tackle, electrician from Butte working on the smelter, who later became a professional boxer under the name "Montana Jack". Three members of the squad who played regularly the previous season were Jake Rentz out of Milwaukee, center; Bernier, boilermaker at the foundry, end; and Barney Fitzpatrick, smelterworker, halfback.

The captain of the team and fullback was Ben W. Wilson, 6'1", 195 pounds. He was a protege of famous coach "Pop" Warner and had distinguished himself as a star with Cornell University in 1897. Wilson had come to Butte to play football but was persuaded to take a job at the smelter general office and to organize the Anaconda squad. He was assisted by Bill Keller, who had played center at the University of Nebraska. Keller played tackle on the new Anaconda team. The team's financial manager, who also played end and quarterback, was R.R. Kilroy, principal of the Anaconda High School. He had been a star rugby player at Trinity College in Dublin and a cowboy in eastern Montana before coming to Anaconda. Kilroy later became the editor of the *Anaconda Standard* in 1919.

As Anaconda formed its new team, Butte discovered that it had more than enough talent for two "big teams", many of whom had been attracted to Butte by its football fame. Jim Hooper and Perry

Benson, both local players each organized a team. Benson's team was called the Montana Athletic Club, but locally the teams were known as the Hoopers and the Bensons. In the meantime the rules of the game had been changed and a touchdown now counted five points rather than four.

The season opened in Butte on October 15, with Anaconda playing the Hoopers. It was a snowy day with six inches of wet snow on the field. The Hoopers beat Anaconda by one point, seven to six. The next week on October 22, Anaconda beat the Bensons ten to five. On November 5, Anaconda fought the Hoopers to a draw, nothing, nothing. This turned out to be the last game of the season as the remainder of the schedule was canceled because of financial problems.[10]

It was also the last year of the "big teams" in both cities. After that the inter-city football rivalry lived on over the years in a series of independent football leagues in which Anaconda was represented by at least one team, usually called the Anodes, made up entirely of local boys. Those teams won enough championship trophies to earn respect in Butte and general recognition throughout the state. And for those natives of the "old sod" not satisfied with "such sissy tussles", there was the long standing Gaelic football rivalry between the Wolftones of Butte and the Emeralds of Anaconda, begun before the turn of the century and played regularly until 1926.[11]

The coming of high school football to Butte and Anaconda closely paralleled that of the "big teams". Butte had its first high school team in 1893, Anaconda in 1898. The high school rivalry dates from that year with Anaconda beating Butte in its first encounter and establishing its credentials as a serious contender in Montana high school football.[12]

Over the years, it was in the realm of high school sports, not just football, that Anaconda demonstrated its fierce pride and its will to win. Since Butte had four times the population of Anaconda, Butte high school was four times larger and had four times more athletic talent to draw on for its teams. But though Butte high's larger and stronger teams won many more games than they lost to Anaconda, they could never take the smelter city for granted. Butte high always had to practice hard and play its best in order to win because Anaconda won with enough frequency to be considered a permanent threat. Anaconda high, for its part, thrived on this rivalry always looking forward to its annual encounters with Butte and reveling in its occasional wins.

This spirit, imbued by Daly, that Anaconda was a winner and would always take its place among the front ranks of Montana cities, inspired high school football and basketball year in and year out, as the town fielded quality teams against much larger rivals. While it had grown almost overnight into one of the state's leading cities, Anaconda high was never one of the larger schools. Unlike most high schools in the state, which drew their students from large rural counties, Anaconda high's entire student body came from the town and its immediate environs, limiting its size. Yet, the school represented the town and was expected to win against any school in the state. Just as it always played Butte every year, for decades the state's largest high school, Anaconda high scheduled games with Montana's other large high schools, even when it ranked only eleventh or twelfth in size in the state and presumably should have been playing in another league.

Despite the town's successful "big teams" in football, bigtime baseball never came to Anaconda. While records indicate that Anaconda teams played Butte and Deer Lodge from 1886 to 1889, The *Anaconda Standard* editorialized in 1890 that baseball still had not gotten a foothold in Montana and that "Butte has never seen good baseball". But this didn't dampen local enthusiasm for the game or for the organizing of local teams. A few days after that editorial the *Standard* announced that a meeting was to be held at Peckover's cigar store to discuss the organization of a baseball club, and that, weather permitting, a game was to be played after the meeting. It noted that "Anaconda was well represented last year with a first-class club that could hold its own with any club in the state. A fine ball ground was fitted up at the driving park" so expenses would be minimal. It declared that while Anaconda did not aspire to belong to join the Pacific Coast Association it could play teams from Helena, Dillon, Butte and Marysville. The April 12, 1890 *Standard* noted that there were youngsters practicing baseball at the field behind the depot and that Anaconda needed a local juvenile league.[13]

Teams were organized that year and Butte and Helena made an effort to get into the Northern Pacific League. There were also suggestions for forming a Montana League with eight teams from

Butte, Helena, Granite, Anaconda, Philipsburg, Missoula, Great Falls and Dillon.[14]

Subsequently Butte did get into the Northern Pacific League and Anaconda had one or two teams each year which scheduled games with other teams around the state. Playing fields were constructed at the race track and the city park in Anaconda as well as the field behind the depot and at the athletic park in Carroll. During the summers, baseball became a passion among the younger men of the town. By the turn of the century, a junior baseball league had grown from just a few teams to over thirty. Some of these teams were to graduate players who went on to have careers in professional baseball.

On one makeshift field in the middle of town, two games a day were played every day during the season. Between games, young players could be seen out on the field raking rocks and preparing it for the next game. The popularity of these games prompted the Anaconda Company to grade and recondition the field, expand it to a whole city block and build a fence around it. This was the origin of the City Common, a center for diverse outdoor activities in addition to baseball and the closest Anaconda ever came to having a town Square.[15]

High school basketball and track were also popular, and in contrast to the other athletic activities of the city, women were encouraged to participate. High schools were represented by both men's and women's teams, and games were scheduled accordingly. In 1906, Anaconda High School men's team won the state track meet by a wide margin as a result of the athletic prowess of a young man named Joe Horn who took honors in every event he entered. That same year Dan Sullivan from Anaconda, a member of the U.S. wrestling team, came close to taking the gold at the Olympics in Athens. An article in the *New York World* claimed Sullivan would have won, but that the judge had coached and then unfairly favored the Finlander who took the medal.[16]

Interest in Canadian lacrosse spread to Anaconda in 1890, but died out soon after. The first formal game played in the city took place at the Anaconda Butte Hibernian picnic at the city park. The August 10, *Anaconda Standard* had this to say about the game, "Today at the Anaconda park will be played the first match game of Lacrosse ever played in Montana....Butte has a good team, some of the best men in the country have been secured by the managers of the team. Many Anaconda boys have never handled a lacrosse stick until a few weeks ago, so the Smelter City lads are not in as good shape as they will be by the close of the season. But good enough to make the contest interesting and perhaps exciting." For an inexperienced team, Anaconda, did surprisingly well. It held Butte scoreless for forty-five minutes when the game was called because of rain, score 0,0. But the sport never caught on and the following year neither town fielded a lacrosse team.[17]

In addition to organized sports, some of the main attractions at holiday celebrations and picnics were contests of skill and strength. On June 13, Butte Miners Day, the miners always put on drilling contests, which consisted of two-man drilling teams swinging large hammers to drive a pointed steel bar (drill) into a solid block of granite. The team that could make a ten-inch hole the fastest won. The tug-of-war was another sure crowd-pleaser at these celebrations. The strongest and the heaviest men were recruited for these events and some of them took it seriously enough to engage in regular practice sessions beforehand. The newspapers published reports about team preparations and played up the rivalries by encouraging charges and counter-charges from various team members. Before the smelter-mans' picnic one year a reporter interviewed Jim O'Brien, the leader of one of the Butte teams, who had been heard boasting that his team didn't need any practice to beat the best of Anaconda. When asked about it, he said in a heavy Irish brogue:

"Ah, ya got us wrong, lad. Sure we're gonna practice. It's just that it's so expensive."

"Expensive?" asked the puzzled reporter.

"The ropes ya'know. They cost a lot of money and me lads keep breakin' 'em." After a long pause to let his full meaning sink in, he continued, "But we're gettin' a steel cable that were gonna cover with rope. As soon as that's ready, we'll start practicin'."

One of the most serious and long-remembered incidents growing out of a such a contest occurred at Gregson Hot Springs during a Butte Miners' Day picnic when the Anaconda Mill and Smeltermens Union tug-o-war team beat the Miners' Union team. Since it was a picnic, whiskey and beer flowed freely, intensifying the excitability and lowering the boiling points of participants and spectators alike. As soon as the match was over, loud arguments led to traded punches and then total chaos as all the members of both unions waded into the fray. It's said that the hospitals of both cities were full for the next couple of days and that the company's production figures suffered a precipitous, if temporary, decline.[18]

But of all recreational activities, the two that had by far the largest number of participants year after year were hunting and fishing. This was after all a frontier town in a sparsely populated state, which had extensive forests and large wilderness areas where big game was plentiful and mountain rivers, streams and lakes held an abundance of fish. Some big game animals like moose, elk, deer, bighorn sheep and bear were hunted for meat. Others like wolves and coyotes were hunted as predators. Smaller animals such as rabbits, weasels, badgers and beaver were sometimes hunted for their fur, but more often for sport as were ducks, geese, pheasants and grouse. Almost every household in town had a number of fishing rods and at least one rifle or shotgun and a day off from work was an opportunity to get out of town to hunt or fish.

With so many gun owners it is not surprising that there was also a live interest in shooting competitions. Both Anaconda and Butte had gun clubs and competed regularly in competitions organized locally as well as at the more prestigious ones organized by firearms and munitions companies. The Winchester Arms Company and the Dupont Powder Company and other such organizations employed professional marksmen to compete in these contests and promote their companies' products. Both cities were also frequent sites of state and regional competitions. And members of the Anaconda Gun Club traveled to meets in other cities in the West as well as competing locally.

During the long winter months, dancing, skating, sledding and icefishing were popular. There was at least one dance or ball a week organized by one or another of the many orders, lodges, guilds and societies in town and held in one of the local halls. Local orchestras would actively promote dances, as well, to drum up business for themselves.

Sleigh rides were sometimes organized in conjunction with dances at recreational spots outside of town. On a bright, cold, winter evening large sleighs filled with straw and buffalo robes, pulled by four to six horses, each led by a man on horseback carrying a lantern would gather merry-makers from various parts of town and then move on to Gregson Hot Springs, the Warm Springs Hotel or the Three Mile House for an evening of dancing to piano and fiddle music. There were tales of late night return trips to town when pranksters among the passengers would pay the drivers to let their teams go off the road and overturn the sleighs into a snowy ditch or a snow bank to add to the adventure of the evening.[19]

Walnut Street and Birch Street, both with good slopes, became favorite sledding and coasting areas, and it wasn't long before serious bob-sledding became a popular activity on Birch Hill throughout the winter. The bob-sled route was an icy four-foot wide path from the top of Eighth Street, down the middle of Birch Street, across the width of the town, and over the BA&P tracks. The sleds came by so fast, up to a mile a minute as they passed fourth street, that citizens on routine errands approached Birch Street with extreme caution in the winter time. During the daytime, mostly youngsters used the slides; but in the evenings, the big bobs took over. There were sleds of many sizes that could accommodate from four to sixteen riders. The largest ones were constructed specifically for the Birch Street run. The most famous of them all, built by the pattern shop at the foundry, was even outfitted with a headlight. Some weighed up to one hundred and fifty pounds. Each had a name. The one with the headlight was "The Foundry Bob". Others were "The Drunkard's Dream", "Lightening Rocket", "The Rambler", "The Wabash" and "The Special".

A major hazard for the bob-sleds were the street cars, so signals were rigged to indicate a clear run. Nevertheless, there were accidents and upsets from time-to-time and accompanying injuries. Mothers worried about their youngsters getting hurt and extracted reluctant promises from them that they would not go near the slide on Birch Street and they wouldn't ever ride the big bobs. But the attraction and the fun often proved irresistible. Sledding down Birch Hill on a big bob-sled was an experience not to be missed and a regular part of winter in Anaconda until the early 1930s when automobiles became too numerous, accidents too frequent and the danger too great.[20]

Another favorite winter attraction was the Scottish sport of curling, which became popular in western Montana in the 1890's. This is a game played on ice with large, specially rounded granite stones with handles on them. The stones are propelled over the ice toward a large circle called a house. There are four players on each team with two stones apiece. The object of the game is to curl the stones as close to the tee in the center of the circle as possible and to knock the opponents' stones out of the circle. Four curling rinks were built in Anaconda around the turn of the century and there was lively competition over many winters not only between local teams but with teams from Butte, Missoula and Great Falls as well. In 1919, Anaconda hosted the state championship bonspeil in which thirty teams participated. Curling continued to be a regular part of the winter scene until the 1930's when the local clubhouse burned down. Although sporadic games were played after that, the sport died out.[21]

While ice skating was a regular winter activity in Anaconda, hockey never really caught on. Year after year, teams would be formed to play the better organized Butte clubs, but they would disband after a few disappointing performances. Unlike the long-lasting curling clubs no permanent hockey club was ever organized. In skating Butte clearly had the edge, producing not only good hockey teams but many outstanding speed and fancy skaters over the years.

9. Growing Up

In addition to sports, social life in Anaconda revolved around civic, fraternal, social, religious, labor and political organizations. Alexis de Tocqueville, in his "Democracy in America" says, "The Americans make associations to give entertainments, to found establishments for education, to build inns, to construct churches, to diffuse books, to send missionaries to the antipodes...If it be proposed to advance some truth, or to foster some feeling...they form a society."[1]

And so it was in Anaconda. These organizations not only fulfilled practical and utilitarian needs, they were also the center of social life for a good part of the population. Anaconda was no different from any other small American town in that regard. Many of the organizations came into being almost simultaneously with the town itself and defined to some extent the prevailing and conflicting influences alive in it.

The Irish from Butte and Walkerville were quick to help their countrymen in Anaconda start a unit of the Ancient Order of Hibernians. When Division 1 for Deer Lodge County was formed September 13, 1885, with Thomas Daly, President; and Mike O'Rourke, Vice President, it immediately became active in the city's public events, organizing social evenings, dances and marching in parades. The unit's 1888 minutes indicate that seventy-five of its members marched in the July 4, parade following the Marcus Daly Engine and Hose Company. On Thanksgiving eve, 1892, the Hibernians sponsored a grand ball in the Evans Opera House. And in 1899, with a loan from fellow Hibernian D.J. Hennessy of Butte, they built a handsome two-story building that rivaled in grace the new city hall which was then under construction across the street.[2]

A few years after the Hibernians organized, natives of Croatia and Slovenia started a mutual aid organization to care for their countrymen in distress, providing assistance during sicknesses and paying for burials. This was the beginning of one of the most prominent and long-lasting fraternal organizations in Anaconda. Since the first meeting was held on the Catholic feast day of St. Peter and Paul, June 29, 1888, it was called the Saint Peter and Paul Fraternal Society. As might be inferred from this name, many of its members were also active in church affairs. One of the founders, George Barich, was instrumental in the building of the town's two Catholic churches. Also, members of the Society were listed as trustees of St. Peter's Church, formally dedicated in 1898. Others, besides Barich, prominent in the organization were Mike Mogus, Joseph Sladich, John Barkovich, Mike Herbolich, John D. Rom, John Shutte and Mike Kracker. Like the Hibernians, they built their own hall, sponsored social events and participated in parades.[3]

The Anaconda Acacia Lodge of the Ancient Free and Accepted Masons, was started in 1886, and the Benevolent and Protective Order Of Elks established a lodge in Anaconda in 1892. Other organizations begun about the same time were the Independent Order of Odd Fellows and the Knights of Pythias. In 1886 Union army veterans from the Civil War also established a chapter of the Grand Army of the Republic. In the following years, a host more of other groups were founded such as Woodmen of the World, Modern Woodmen of America, Order of Pocahontas, Pioneers of the Pacific, Catholic Order of Foresters, Independent Order of Foresters, and the Improved Order of Red Men. In 1898, forty-five women assembled in the Montana Hotel to found the Anaconda Women's Literary Club. Mrs. Phoebe Hearst, who addressed the group on June 11, 1898 was made the club's first honorary member.[4]

After the turn of the century, the list gets still longer and includes St. Ann's Society, St. Philip's & Jacob's Society, the Ancient Order of United Workmen, the Brotherhood of American Yeomen, the Fraternal Order of Eagles, the Order of Scottish Clans, the Royal Highlanders, the Scandinavian Brotherhood, the Knights of Columbus, the Cristoforo Columbo Society and the United Artisans. A small group of German businessmen optimistically formed the short-lived Twentieth Century Club. The Swedes established a chapter of the Order of Vasa and the Norwegians the Sons of Norway. Newly arriving Eastern Europeans formed the Nardon Slovenski Spolak National Union and the Benevolent Society of St. George, which were eventually absorbed by the St. Peter and Paul Society. Like their predecessors, these organizations offered fellowship and support. They all organized social events, sponsored fund-raising activities, participated in parades and other civic functions and provided assistance, hospital care and death benefits to individual members. Many had their own marching bands.

Some of the Irish were also members of what were known as military companies, formed according to Hugh O'Daly "for the purpose of taking part in any revolution that might take place against British rule in Ireland"[5] As a way to defeat the hated oppressor, these companies drilled and trained, ready to join in what they hoped would be the invasion and annexation of Canada. William M. Kelly organized the Wolf Tone Guards Military Company of Anaconda in 1888. There were also the Hibernian Rifles and Shield's Guards. While these valiant Irish patriots were never unfortunate enough to have to prove themselves on the battlefield, their organizations made handsome additions to local parades with high spirits and smart marching bands, as well as sponsoring annual balls and numerous other social events.[6]

Typical of such activities was the 1890 annual Anaconda Butte Hibernian picnic where Hibernian Guards of Butte, the Thomas Meagher Guards of Helena and the Wolf Tone Guards of Anaconda paraded from the train depot to the Montana Hotel and performed the manual of arms before marching in formation to the park. The Anaconda Hibernians presented a prize sword to the Thomas Meagher guards for the best performance.[7]

By the time the Montana Hotel and the race track were built, Daly had already moved his wife and four children from Butte into a large new three-story brick house on the corner of Sixth and Hickory Streets. It had a ball-room on the third floor and a large two-story brick barn in the back for the family's coach horses, with an apartment above for their coachman. Daly brought other members of his and his wife's family to Anaconda and built houses for them, as well. Two sisters had houses around the corner from his; his brother and father-in-law also had houses in the same neighborhood.

Daly's wife, Margaret, immediately began taking an active interest in the town. She worked closely with the Reverend W. E. Nies to plan the new St. Mark's Episcopal Church on Sixth and Main Street in 1890. Modeled after a typical English village church, it was constructed of field stone in a cruciform. The cornerstone was laid on October 21, 1890. All the stone for the church came from Garrison, Montana. Once the church was completed, Mrs. Daly donated a twenty-two rank Steers and Sons pipe-organ, the largest then in the state. This was the same year that Reverend Philip Lowery completed his Methodist church on the corner of Oak and Third street.[8]

In January 1891, the Anaconda Mining Company became a corporation, with a capital of 500,000 shares at a par value of $25 a share. The new smelter, or lower works, was now in full production handling 3,000 tons of ore a day. Twelve converters produced 225 tons of blister copper every 24 hours. In 1892, the upper and lower works together produced over 100 million pounds of blister copper. This increase lifted the United States into first place in world copper production, demonstrating America's growing industrial power, and making Anaconda the largest producer of copper in the world.[9]

The new village of Carroll had become a bustling community containing newcomers from Michigan copper country and many immigrant Irish, most of them single. There were more smeltermen living in Carroll than in Anaconda. Before 1889 the only public transportation between the two towns were horse-drawn three-seated surreys, that could carry about nine passengers each. In that year, construction was begun on a street car line which would run from the race track west of Anaconda through the town and east to Carroll and the lower works. The line was laid down Third Street to Main where it turned north past the Leland and Montana hotels, past Park to Commercial Street where it turned and continued east to Cedar, back south to Sixth and east to the Foundry, then northeast across the flats to Carroll and about a mile further east to the Carroll Athletic Park. To provide electricity for the new trolley line, a dam and power-house* were built west of town on Warm Springs Creek. The new powerhouse was equipped with ten 500-volt hydroelectric generators. It would also provide power for the hotel, the *Anaconda Standard*, street lights and parts of the smelter. The completely electrified street car line with single-truck trolleys was inaugurated in September 1890.[10]

It is interesting to note here that Anaconda had hydroelectric power only eight years after Thomas Edison had built the first hydro-electric plant in the United States at Appleton, Wisconsin and only four years after the first hydroelectric generators were installed at

---

* Myers Dam, Holmes Powerhouse.

Niagara Falls. And its electric trolley system begun only two years after the first one in the United States was inaugurated in Richmond, Virginia.   A new company called The Electric Light, Power and Street Railway Company was formed to manage the power house and the trolley line.

Anaconda narrowly missed being the first town in the state to have an electric street car system. Helena inaugurated its system two months before and Butte one month before Anaconda. The street car became Anaconda's primary means of public transportation, not only for workmen going to work, but for housewives going shopping and families attending sporting or social events. Nevertheless, horses and horse-drawn wagons were still much in evidence on city streets.

With changing needs over the years, the line was rerouted several times and the single-truck trolleys were replaced by larger motor cars seating 60 people and able to pull 5 additional trailer cars, which seated 40 passengers each. From 1902 on, they usually consisted of a lead car, with its long trolleys sucking juice from the overhead wire and spitting out sparks, pulling four or five trailers. Before then, the smaller trolleys pulled only two or three trailers of open cars. These had long board steps running the length of each car and benches facing each other across their width.[11]

When Daly's third enlarged smelter was put into operation on the southern hills across the valley in 1902, the lower works was closed down and Carroll was dismantled. The trolley line was then rerouted straight down Third street and east up smelter hill. The turns down Main Street and Commercial, up Cedar and down Sixth were eliminated. A short spur was added from Spruce Street over to the dance pavilion in the newly expanded city park.

Transportation was a major issue for Daly during these years, but not only to get his smeltermen to work. He was still unhappy with the deficient railroad service between Butte and Anaconda and annoyed by the high freight charges for ore shipments by the Montana Union railroad, owned jointly by the Union Pacific and the Northern Pacific lines. Competition on the international copper market was fierce and high rail charges that raised production costs put Anaconda at a disadvantage.

In 1891, when it came time for Daly to renegotiate his contract with the Montana Union, he wanted the rates for ore shipments reduced from seventy-five to fifty cents a ton. When the railroad

6. Five-unit electric street car trains making their way up smelter hill were a common sight after the Washoe Smelter was completed. When the eight-hour day was instituted in 1906, there were two such trains for each shift arriving before 7:00 a.m., 3:00 p.m. and 11:00 p.m. and returning those who had completed their shifts. There was an additional train for day workers which arrived before 8:00 a.m. and returned after 4:00 p.m.

management refused to make this change, Daly, in desperation, shut down his mines in Butte and the smelter in Anaconda, throwing over 2,500 men out of work. The ensuing tug-of-war lasted five months. Daly eventually got the lower rates he wanted, but the expense in lost production was so high that he vowed to build his own railroad.

He contacted Jim Hill of the Great Northern, who already had a contract with Daly for shipping blister copper from Butte to a new refining plant the Anaconda company had built in Great Falls. The Great Northern bought the bonds which Daly floated to build the new line and Hill assigned Frederick Whythe, from his staff, to help Daly survey and plan the route which would parallel the Montana Union tracks through Silver Bow Canyon. As with everything else Daly did, his railroad was to be an endeavor of consequence. He called it the Butte, Anaconda and Pacific; it would run from Butte through Anaconda, west into the Bitter Root valley to Hamilton and on to San Francisco. The BA&P was incorporated in October, 1892 and the

Butte-Anaconda portion began service on December 27, 1893. While intermittent building on the line continued after that, the furthest the railroad ever extended was twenty miles west to Georgetown and the Southern Cross mine. This section was built in 1911 and 1912 when the Anaconda Company took a renewed interest in the gold ore at Southern Cross.

From the beginning, the Great Northern treated the passenger service on the Butte-Anaconda run as an extension of its own Montana Central Railway from Great Falls to Butte and operated sleepers from Anaconda to St. Paul and on to Chicago. An advertisement run regularly in the *Anaconda Standard* in 1893 stated: "the Great Northern makes quicker time from Butte to St. Paul than any other line and is the only road which makes the run to Chicago in two nights..." This passenger arrangement between the BA&P and the Great Northern continued until about 1904.[12]

Like other industrial magnates of the time, Daly had a handsome private railroad car built for himself. His car could often be seen attached to the end of an ore train forty or fifty cars long wending its way slowly through the canyon to Anaconda or returning to Butte.

Daly's new railroad achieved his primary objective. As many as four trains a day ran between Butte and Anaconda, dropping the average cost of transporting copper ore during 1894 to twenty-five cents a ton, one-third of what the Montana Union had been charging. The line also stimulated a lively passenger service. During the horse-racing season, a daily train of up to fifteen coaches left Butte at one o'clock filled with race-goers for Anaconda. Street cars met this train at the Depot on Main and Front Streets and took them to the race-track. After the races, people left the track for a ride downtown or back to the train station at a special loading platform for street cars. Later the railroad built a spur directly to the race track.

By the time Anaconda was seven years old in 1890, there could be no doubt that this was not just another flash-in-the-pan future ghost town. It had a population of almost 6,000 with 1,160 men on the smelter's payroll. Already it had permanence and position in the state, and the pace of new building was accelerating. Prominent citizens in the town such as J.H. Durston, and C.W. Tuttle were building large homes. Conrad Kohrs from Deer Lodge built a large house on west Third Street. The number of telephones in town had risen from 18 to 28. And it was already bragging that it had the finest volunteer fire department in the country, with four companies of 175 men, 14 electric fire-alarm boxes, and a firehouse tower 58 feet high containing a watch room and a bell donated by Marcus Daly.[13]

Several reservoirs had been built west of town to provide water for both the smelter and the town. Businesses and most of the residential property owners took advantage and had water piped in from the water mains, thus even before the turn of the century, most residents of Anaconda enjoyed the luxury of running water. In addition, six cisterns, each holding several hundred gallons, were built at strategic locations in the downtown area for fire fighting. In case of a fire, water from these cisterns was turned into ditches which ran down the main business streets. The fire companies first relied on hose carts and hand pumps. Later horse-drawn fire engines and steam engines were used to pump water from the cisterns.[14]

Most of the boarding houses and saloons were concentrated in four block area along East Front Street. That made them not only convenient to the train station, but also within walking distance of the Upper Works via the main road which crossed Warm Springs Creek at Cedar Street. In keeping with the spirit of civic betterment, many of these establishments sprouted new names. Jim Keefe's, still on the corner of Front and Main, was no longer called the Alamo but had become the St. James Hotel, and others along the street followed suit. John William's place became the California House, Joe Mulvihill's the Del Monico, Con O'Connor's the Pacific and Mrs. Shultz's, the Grand Hotel. Not to be outdone Mrs. Ramsey called her hostelry the Grand Central and A.J. Blix advertised his Atlantean prominently in the *Standard*. There was also the Saratoga House and the Merrimac.

Most of the stores and businesses were one block over on east First (later Commercial) street. This is where Sam Pramenko had his grocery store, with tables of fresh fruit and vegetables, out on the sidewalk. He had grapes packed in sawdust, live chickens in cages, a barrel of oysters shipped in from the east, and sacks of potatoes and flour. J.P. Dunn and A. Mandoli also had grocery stores here and George Barich his feed store. David Cohen had a general merchandise store on the corner of Commercial and Cherry with long benches in front loaded down with shoes, socks, caps, hats, piles of pants, shirts, lunch buckets, candles and anything else that would sell.

David Walker and Nick Beilenburg had a butcher shop on Main near First with a half-side of beef or pig hanging outside the door and inside behind the counter freshly killed chickens hanging by their feet. On the opposite corner was the Arcade Saloon, "An elegant free lunch served day and night". Further down First was Neal & Son Newsdealers and Confectioners, then A. T. Playter's drug store. Dr. St. Jean, who was in attendance at St. Anne's Hospital, also had offices over the drugstore. Next door was a dairy with milk cans sitting beside the door and large wheels of cheese and lumps of butter wrapped in cheese cloth on tables. On the second block down one could get a whiff of the rich smell of fresh bread and sweet rolls coming from Schroeder's Viennese Bakery and, just across the street, Ben Falk's butcher shop.[15]

First Street had most of the livery stables and blacksmith shops, as well. Warm Springs Livery, on First just west of Main Street, advertised its "fine new hearse". Just east of Main street was D. B. Birren, Carriage Builder, Blacksmith and Wagon shop. On the block between Cherry and Cedar, Sawyer, Houch & Co. sold wagons and carriages. On the same block, E. Jacobson sold furniture, stoves, cookery and funeral goods out of his store and ran an undertaking service on the same premises. Across the street W.C. Haynes ran a livery stable and on the corner was a butcher shop, called Peoples Market, which sold both wholesale and retail. In the next block J. B. Gnose had a general store and there was another wholesale and retail butcher shop run by W.P. Burrows named the Central Market.

The post office, originally on First Street had moved to Main Street, across from the Montana Hotel. Also along Main were Read's Drug store on the Corner of Front Street, I.F. Kirby, "Hardware, Tinware, Glassware", the Del Monico hotel, Joseph Murray's cigar and candy shop, Losee & Maxwell Women's Wear next to the post office, Barret & Jacky's Buggies, Carriages and Harnesses, J.L. Hamilton "Fancy Groceries and Provisions" opposite the Evans Opera House, and J. Ross Clark Groceries on the corner of First and Main.

In 1889 M.S. Aschheim had a grand opening for his "Mammoth Dry Goods Emporium" on First Street, and in 1890 the handsome DeLaurier building replaced an old blacksmith shop on the corner of First and Cedar, and McKinnon and McKay opened a large grocery store in the Hoff block on the same street not far from D.G. Bownwell's Anaconda Livery Stable.[16]

7. Copper City Commercial Co., the company store. The largest and most diversified store in town. Before the turn of the century smelter workers were expected to do all their shopping here. The store closed after the 1921 recession when the smelter was shut down for eleven months.

This was Marcus Daly's town, owned and controlled by him and those around him. His company owned and operated the smelter, the railroad, the bank, the newspaper, the main hotel, the race track, the street car line, the power company, the foundry company, the firebrick company and a number of coal and timber companies providing wood and coal to both Anaconda and Butte.

In 1891, J. Ross Clark, W.A. Clark's brother and Daly's brother-in-law, and Daly's good friend, D.J. Hennessy, who had a store on the corner of Oak and First Streets and ran mercantile stores in Butte and Missoula as well, proposed opening a new store in Anaconda. This store would be jointly owned by them and the Company and would supply all the needs of company employees and laborers.

Thus, the Copper City Commercial Company was formed and opened its doors to the public in the early spring of 1892. It was a large enterprise, housed in a two story brick building, which took up half the block, on the southwest of Main and First, where J. Ross Clark's store had been, with warehouses across the alley in another building that fronted on Hickory street. Its stables and wagon sheds were across the street on the northwest corner of Main and

First, with a yard large enough to accommodate the teams and wagons of farmers in town to buy or sell. The new establishment caused a sensation when it opened, not unlike the opening of a modern day shopping mall.

It also caused concern among other Anaconda merchants who feared unfair competition. Estes and Connell, the largest mercantile store in town, didn't even try to compete, it sold all its stock to the new enterprise and went out of business in April, 1892. The only other large mercantile left was McCallum & Cloutier, which managed to survive and even to out-live the new Copper City and its initial built-in advantages. McCallum, who like his partner Cloutier had come from Canada, began his business career in Anaconda as clerk and bookkeeper in Sam Pramenko's store. Sam had helped him open his own store by vouching for his credit to obtain his start-up stock.[17]

Young Spenser L. Tripp, who worked as a clerk in the McKinnon & McKay grocery store on First Street wrote the following in a letter dated February 7, 1892: "A half-dozen of the leading men, employees of the Company, have formed a Company for the purpose of controlling the general trade of the town. The promoters are all Irish Catholic and will control the trade of their own breed, besides making it hot for any white man holding a position with the Company that does not patronize them. I am sorry to say that the town is getting to be thoroughly 'Mick'." And this on April 17, 1892: "...The new outfit has taken over every single Irish customer that we had leaving us a small minority of family trade—that is the American and English families with a few Swedes. The average flannel mouth loves to talk about freedom, but when the opportunity arrives that enables him to put on the thumb screws he does not hesitate a moment but goes to work forgetting his previous protestations." [18]

With a captive market, each of the Copper City's departments was larger than any of the specialty stores in town. It had a large and well-stocked grocery department with meat market, a notions and dry good store, a men's and boy's clothing department, a large shoe department for both males and females, a ladies ready-to-wear and millinery department and a huge furniture and hardware department. Jack White, whose father worked there remembers, "upstairs during Christmas season was the largest toy department that I have ever seen".[19]

Matt Kelly says: "It was understood and made quite plain to the smelter workers that they were expected to trade at the company store...Mr. Lyman, one of the solicitors, would go through the Works every day. He would ask a new man how he liked his job...If the man did and wanted to hold it, Lyman suggested that he had better trade at the Copper City, and would give him a charge book. All payrolls were sent to the Copper City for the smelter, foundry and railway employees, and the amount of their unpaid accounts held out...

"Paydays came once each month, the 10th, 11th, and 12th for the various departments. The women generally called for the pay as the men worked such long shifts that they couldn't get to the store until late. On pay days, the line of women extended from the office out through the shoe department and to the front door, all day long. The men lined up through another department into the office." Pay was in gold and silver; gold in five, ten and twenty dollar pieces and silver in dollars or in one bit (12½ cents), two bit (quarter) and four bit (half dollar) pieces.[20]

The Copper City didn't limit its coercion to individual wage earners, either. Independent merchants were cajoled and encouraged to establish accounts at the store not only to buy needed supplies but to turn over bill collecting, for a fee, to Copper City, which administered Daly's payroll. It took a strong willed merchant to resist their efforts. The family of Joseph Elie, claims that he closed up his shoe-making shop and his "Bonanza" shoe store and left town when he was pressured to trade at the company store. He left even though he made shoes for Marcus Daly.[21]

In spite of such unfair practices, young Spenser Tripp quit McKinnon and McKay and with a friend, Rob Dragstedt, opened a store in 1896 and prospered for several years until the Anaconda Company began paying part of the workers' wages in script which could be used only at the company store. In 1899, no longer able to withstand Copper City's advantage, Tripp and Dragstedt closed their Anaconda store and moved to Butte. There the business grew into one of the largest in the state.[22]

Being a large establishment, the Copper City had a large staff of people working for it including managers, bookkeepers, clerks, buyers, warehousemen, grocers, meatcutters, butchers, dry-goods salesmen, teamsters and delivery men. Some of the staff was brought in from outside and others were hired locally. A number of local merchants closed their businesses and went to work for Copper City. Later, some who had worked for Copper City and learned the trade opened their own stores. E.E. Moore was the general manager for years, other managers were Joseph Peters, George Tighe, Shelly Tuttle and Joseph Kelly. In 1915 Kelly established a grocery business and meat market of his own that continued to operate long after Copper City went out of business following an eleven month shutdown of the smelter in 1921.[23]

10. Chinese Boycott

The same inescapable combination of gold and railroads which settled the West brought the Chinese to Montana. When the golden spike was driven on first transcontinental railroad at Promontory Point, Utah, in 1869, 10,000 Chinese who had been employed by the Central Pacific Railroad laying track, suddenly lost their jobs. Instead of returning to California, where they'd been hired, many of them moved into mining camps in Montana, Utah and Idaho. Because of this sudden influx and later widespread unemployment caused by the Panic of 1873, a vocal anti-Chinese movement, started in California, spread throughout the West and persisted, in varying degrees of intensity, for the next fifty years.

The first sizable immigration of Chinese into the United States came during the California gold rush, when America became known to thousands of poor immigrants from south China as "The Mountain of Gold". The Chinese, like the Irish and many other immigrants who caught gold fever, moved from one new discovery to another, becoming ubiquitous at gold and silver mining sites throughout the West, including Montana in the 1860's.[1]

According to some estimates there were about eight hundred Chinese in Montana in 1869. The 1870 census counted 1,949, which was approximately ten percent of the total population. A federal study of the same date reported "some 2000 to 3000 Chinese domiciled in the Territory of Montana". Because they were discriminated against, they usually ended up working claims that whites had abandoned. Or, in many cases, they went to work for whites in mining camps and on ranches doing "women's work", cooking, cleaning, washing clothes etc. at starvation wages.[2]

The 1880 census listed 710 Chinese in Deer Lodge County (Butte included), 359 in Lewis and Clark County, 265 in Madison and 149 in Missoula. The arrival of the Northern Pacific railroad from the Pacific coast into Montana in 1882-83 also brought additional Chinese into the territory since as many as 15,000 were employed building that section of the new railroad. So by 1890, according to the census the Chinese population in Montana had increased to 2,532. An 1889 article in the *Anaconda Standard,* reported that Anaconda's Chinatown, which occupied both sides of Birch Street between Park and Commercial, had a population of between three and four hundred. A later account, probably exaggerated, estimated that one-quarter of the town's population of about 3000 was Chinese in 1887.[3]

The *Standard* article went on to say that besides "doing about all the washing, all the cooking and a considerable share of the table waiting, bed making and dish cleaning..." they are also getting into trade and merchandise. "On Front Street, Yung Lee has established himself as a boot and shoe artist, and [has] a fine brick store on 2nd (Park) Street, Tuck Hing & Co. announce themselves as fashionable merchant tailors [with the] latest New York fashion plates to which he directs his customers' attention...Mr. Hing is not at all backward in praising the superiority of his cloth and the nicety of his fits." He boasted that American tailors couldn't compete with him since he could sell a suit for $27.00. The same suit would cost $39.00 anywhere else. The Chinese also had garden plots on the outskirts of town, sold produce and ran restaurants, known as noodle parlors.[4]

The first Chinese immigrants, who arrived in the U.S. at the port of San Francisco, formed a brotherhood for mutual support and protection called the Six Companies. All arriving Chinese were received by the brotherhood and welcomed into the organization. They were told: "You will always find food and shelter here among us...When you have earned money from your diggings or wages, you will pay dues into the company fund...We are Chinese in a land of foreigners. Their ways are different from our ways. Their language is different from our language. Most of them are loud and rough. We are accustomed to an orderly society, but it seems that they are not bound by any rules of conduct. It is best, if possible, to avoid any contact with them. Try not to provoke the foreigners. But you will find they like to provoke us. We are comfortable in our loose cotton jackets and trousers and we are used to going barefooted. They like to wear rough, course clothing with high-laced boots. They cut their hair short, but let it grow on their faces. We wear our hair long and braided and we shave the hair from our faces. Since we all want to return to our homeland, we cannot cut our queues..."[5]

The Chinese of early Montana were described as: "colorful and mysterious,...They wore their hair in braided queues and dressed in loose, baggy trousers, tight blouses of silk or cotton, buttoned from throat down and floppy, heelless slippers. Many of them wore broad-brimmed, "high-binder" hats or tight, silk skull caps."[6]

The profound differences in culture and language between the Chinese and the white inhabitants of the mining camps and other Western towns, led to suspicion and friction from the beginning. Like the white attitude toward the Indians, there was little tolerance for the life style and customs of the Chinese. Already in 1866, a Virginia City newspaper article announced, "Women of Helena declare war against Chinese. Unfair competition between slaves and free people. Chinese wresting wash tubs from local washer women."[7]

By 1871, after the first influx from the railroads, the cry became louder. A Helena paper declared that "The Chinaman's paramount objective in Montana is placer mining. The results are alarming because they own or operate 1/3 of all placers in the territory. Once they arrive they are hired for kitchens, on the ranch, in laundries and mines crowding out other nationalities. There should be legislation prohibiting them from acquiring further claims." [8]

An article in the Deer Lodge *Montanian* said: "Chinese are industrious and frugal, but their habits, religion and manner of living are totally unlike those of our people; they can never assimilate our civilization; they are among us not of us; they send their wealth out of the country, conserve little and do not benefit commerce. They do injury to the laboring class of our own people. The Burlingame Treaty of 1868 between the U.S. and China accorded Chinese all rights and privileges accorded any person in the U.S., except naturalization." But the treaty must be abrogated and they must go because "they are idolaters, their religion teaches them to lie, steal and cheat, they are devoid of all principles of honor and justice. We welcome any immigration which strengthens our country. They bring no wives and form no attachments. They build no schools or churches."[9]

In an editorial in 1872, it again called for the abrogation of the Burlingame treaty and observed that many miners were selfishly opposed to expelling the Chinese because they were making money selling them old mining claims which white men had abandoned and would therefore be worthless without the Chinese.

When the railroads later began hiring Chinese to lay track and for other tasks as well, labor resentment grew, and the Knights of Labor, an early national labor organization, mounted forceful opposition, which eventually resulted in the Chinese Exclusion Act of 1882 and a law in 1885 forbidding the importation of contract labor. In Butte and Anaconda, this agitation resulted in an unwritten understanding between labor and management that Chinese could not go down in the mines, work on the smelter or join the unions.[10]

Sometimes the Knights' anti-Chinese crusade led to violence. It was the cause of a major tragedy in Rock Springs, Wyoming, which, at the time, was supplying coal for the Anaconda smelter. In 1885, armed coal miners claiming that the Union Pacific mine owners favored the Chinese, deliberately pillaged and burned the houses of Chinese laborers and killed at least twenty-eight of them before order was restored. No one was ever prosecuted for these murders.[11]

Just as in Wyoming, the Knights in Montana agitated for elimination of the Chinese. Even though the Exclusion Act had stopped further immigration, there were strong pressures to drive out the Chinese already in the country. While the Knights advocated driving them out by economic boycott, the atmosphere created by their agitation sometimes led to violence. The Anaconda Standard in 1890 reported that a Chinese wash house on Front Street in Anaconda was blown up. In 1891, it reported that a gang of 16 hoodlums attacked and killed three Chinese in Butte. Nine months later, "Two Chinese brutally beaten by a thug. Jim Edwards placed under arrest. The man also had been thieving around Anaconda."[12]

Indeed, sometimes an assault on a Chinaman was considered nothing but good fun, especially if liquor had been flowing freely. The *Anaconda Standard* in 1898 recalled an incident on the Fourth of July 1868 when a Chinese gold-panner was hanged at Silver Bow.

"It was not a judicial execution. It was simply the cool premeditated act of a disheartened, yet patriotic and Fourth of July conscious miner who hanged a Chinaman to a cottonwood tree just for the devilment and in the hopes that it might bring luck."[13]

Matt Kelly, describing the same tendency in Anaconda, says, "Various young men occasionally felt that they had to either hang or kill a Chinaman. Catching one away from Chinatown, they endeavored to carry out their intentions, a few times with nearly fatal results."[14]

As the ten-year term of the 1882 Exclusion act was about to expire, renewed agitation began throughout the West to have it extended and the Chinese question once again became a hot issue. In 1891, as a result of continuing Knights of Labor activity in both Butte and Anaconda, anti-Chinese committees with semi-official status were formed using Missoula as a model where a boycott had been started to drive the Chinese out of town. Jim McHugh, a local businessman, thought that there might even be a way to make some money out of the boycott so he opened a restaurant and called it "White Labor".[15]

The March 3, 1892, *Anaconda Standard* reported a Knights of Labor Notice:

> "Any member found patronizing hotels, restaurants, boarding houses, or laundries that employ Chinese will be fined five dollars and suspended until fine is paid. By Order of L.A. No.3711 Joseph Wright R. S., John Barry M.W."

The Anaconda committee met at Tcitzen's cigar store and its actions and recommendations were regularly published in the *Anaconda Standard* to encourage public support. A boycott of all Chinese businesses in both cities was declared.[16]

Even without the Knights campaign, the public attitude toward the Chinese was largely negative. The Chinese were viewed as different and therefore fit objects of scorn and ridicule. Newspaper articles, in the irreverent style of the times, routinely referred to them in pejorative terms as "chinks", "celestials", "almond-eyed heathens" and "moon-eyed mongolians". Humor of even the most morbid kind was not off limits. Much of the flavor of such cavalier treatment is illustrated in an article of the *Anaconda Standard* dated May 17, 1893. Under the headline: "A CHINAMAN DEPARTED" "A Dose of Poison, A Leaden Bullet and A Slipknot Transported Him", date-lined Dewey's Flat, Big Hole, it told of the decaying corpse of a Chinese found hanging from a tree, which authorities had finally gotten around to investigating several weeks after it was first reported. There was a rope around his neck, a bullet through his heart, a bottle of brandy and a small vial of poison in his pocket and a revolver lying close by. Without clarifying what the official view of the cause of death was, the reporter concluded that the man had taken his own life. He then described how the death had come about. The deceased was a gambler who could borrow no more money. So he got brandy and poison put them in his pocket, climbed the tree, tied one end of the rope to a tree branch and the other around his neck. He then stood on the branch, drank the poison and brandy, shot himself in the heart with the pistol and fell dead from the limb being choked by the rope as he fell.

The "Chinaman" was routinely blamed for any and every problem which had no immediately identifiable culprit. George Crofutt, the editor of the Butte City Directory of 1886, apologizes for a number of typographical errors appearing in its final text and facetiously blames it on John Chinaman with the conclusion that "The Chinaman Must Go." A fire in the red light district in Anaconda was attributed "to the carelessness of a Chinaman." Even favorable comments about the Chinese were full of patronizing references, poking fun at them for their heathen ways.

Also, the moralists among the critics made sure to make much of the Chinese vices of gambling, prostitution and dope. The newspapers regularly reported arrests of Chinese for these activities. The *Standard reporting* on gambling and opium dens in Anaconda's Chinatown, stated that most Chinese smoked opium; that a white man would probably be charged four bits, (5O cents) for a pipefull; that favorite Chinese gambling games were fan-fan, poker and faro, and that a move was now afoot to restrict the Chinese to a single district in the city. The *Standard* in 1890 reported that Anaconda policemen "are daily seen driving a team of mongolians by their queues to prison".[17]

Under the headline, "Opium Joint in the Heart of the City," an article in the *Standard* described how Chinese were using a property on West Galena Street in Butte as an opium den. When it was discovered, the white owner kicked them out and called the police who arrested the manager. Another article around the same time said that Mayor Dwyer of Anaconda was going to crack down on opium joints. Another reported how the Chinese were inveterate gamblers and that Hung Lee's store in Butte had become a social center for gambling and selling opium, and was the headquarters of a notorious tong (Oriental brotherhood organization).[18]

Gambling, of course, was common in all frontier towns and mining camps, but it seemed to take on a sinister meaning when engaged in by the Chinese. Their preference for opium smoking over alcohol was also seen as undermining frontier values. But some Chinese weren't afraid to argue the superiority of opium over alcohol. An article in the Butte *Miner*, quotes Yung Lee, the owner of an opium den in Butte, as telling the judge after his arrest: "Yung Lee sell dleams, velly nice dleams. They pay Yung Lee money, he bling 'em one, two hour dleam...no botha — just lay down, sleep, dleam. Much blettah than spend money for Ilish whiskey and want to kill evlybody. Much blettah dleam — no hurt anybody!"[19]

But it was the slave trade in young oriental women which gave the Chinese their most negative image. Since females were routinely bought and sold in China, some were smuggled into this country and were sold into prostitution. Thus, it was easy for even men of the cloth, such as Methodist minister, the Reverend Dr. Raleigh, who gave a series of lectures in Missoula, Butte and Anaconda, to join the anti-Chinese campaign and advocate complete exclusion until they became civilized.[20]

While the driving force behind the anti-Chinese sentiment in the West was suspicion and deep-seated fear of an alien culture, many of the charges against the Chinese, from cheap labor to prostitution, had some basis in fact. Popular opinion was so uniformly negative that it was not difficult to get a majority in the U.S. Congress to extend the Exclusion Act. The Geary Act of 1892 not only extended all restrictions in force against the Chinese for another ten years, but removed much of their day-to-day legal protection, as well. No bail was permitted them in habeas corpus cases, and all Chinese were required to obtain a certificate of eligibility to remain in the U.S. It was the opinion of many that as a result, the Chinese could no longer get legal redress through the courts. From this came the now familiar phrase about "not having a Chinaman's chance".[21]

However, as intense as the anti-Chinese campaign may have been, there were always some voices of reason and a good measure of cussedness among the people. So while the government might enforce its laws, it was not always an easy matter for local committees to enforce their boycotts. From the beginning some people stood fast for the principle of equal treatment spelled out in the Burlingame Treaty. In Montana, territorial governor Ashley vetoed legislation in 1870 aimed at charging special taxes on Chinese laundries, arguing that "equal justice under law requires that all subjects of China residing in our territory should be taxed as our own citizen are taxed—no

more, no less. Any attempt to evade this just requirement by
'unfriendly legislation' is inconsistent with the dignity and character
of the American government."[22]

With regard to the Geary Act, the Portland Oregonian editorialized that it violated the treaty with China and could subject
Americans in China to the same treatment. The *Ogden Standard,* said
that the part of the law which dealt with deportation was unen-
forceable and should become a dead letter. The *Salt Lake Tribune*
argued that the power of the government to exclude one race could
then be just as easily applied to others such as the French and the
Germans.[23]

But these were minority views. Editorial comment in most
Montana papers was negative and in some, extreme. The Butte *Miner*
suggested that all Chinese should be deported using U.S. Navy ships.
No passenger cabins would be needed. The Chinese could even be
used as ballast. In a story datelined New York City, The *Anaconda
Standard* reported that the Reverend Edward Payson preaching at the
old Presbyterian Church on Canal Street had branded the Exclusion
Act unjust and un-Christian. It headlined the story "THE CHINA-
MAN'S FRIEND", "Eastern Churches Defend Chinese", as if to say,
what do they know about it? The *Great Falls Leader,* complained
that the Treasury Department was not enforcing the law. And the
Great Falls Tribune, commented that while the need for the removal
of the Chinese might still be debated, it was clear that no more
immigration of "that peculiar people" is allowed.[24]

While plenty of people might have thought the Chinese
peculiar, they saw no reason to stop doing business with them. The
Anaconda paper reported: "There's a good looking young lady
waiting table in a prominent restaurant on First Street who persists in
patronizing the heathen Chinese..." and went on to insinuate crudely
that there was a romantic attachment involved. And another one,
"Anti-Chinese Committeeman caught playing cards with a Chinaman." Another reported: "The Chinese commenced celebrating their
New Year, Wednesday Feb. 15, and will keep it up all week. In spite
of the boycott, they're having a good time." It also carried reports
that the Chinese were purposely walking around town with full
baskets of laundry on their heads to advertise that they were still in
business despite the boycott.[25]

The Citizens Anti-Chinese Committee of Anaconda intensified its campaign in early 1893, by publishing a resolution in the *Anaconda Standard* on February 13, and again on February 14, asking citizens to withhold patronage from the Chinese and "resolved that they should not live amongst us."[26]

In February 1893, the same Committee called for a mass meeting at Evans Hall to educate the public. It wanted the names of everyone employing Chinese so the "addresses of the guilty parties could be forwarded to the secretary ". It also directed a message to ranchers in the Valley to begin supplying truck-garden vegetables to replace those of the Chinese. The Deer Lodge County Trades and Labor Council asked that patronage of Chinese businesses be ended.[27]

The Butte Committee was also active. The *Standard* reported that "The Butte Anti-Chinese Committee has taken upon itself to visit Madames on the Row and demand they fire all Chinese." When they made these demands, the Madames ordered their Chinese employees to throw the committeemen down the stairs. The committee members left in a hurry promising to put a boycott on the houses. The Butte Labor and Trades Assembly issued a notice that the names of those employing or patronizing Chinese would be published.[28]

But with time the boycott began to take its toll in Anaconda. As early as January, 1893, the *Anaconda Standard,* editorialized that the boycott was working and warned against brutality. It reported that the Chinese owners of the restuarant in the old Read Building wanted to sell it for one-half of what it cost. Reports such as the following were also appearing in the paper: "Chinese Leaving by the Dozens" "Seeking towns where there is no boycott against them." "Chinese who conduct a vegetable garden down near Tuttle's Foundry want to sell out because of the boycott." "Chinese leaving Portland selling their reentry permits."[29]

While the boycott in Anaconda seemed to have had the desired effect with the departure of a large part of the Chinese community, it was deemed a failure in Butte which still had a substantial Chinese population. As a result another boycott was started there in late 1896 headed by the Cooks and Waiters Assembly, the Hotel and Restaurant Keepers and the proprietors of three steam laundries. The Silver Bow Trades and Labor Assembly, made up of thirty different unions endorsed the boycott in January 1897 and the Butte Chamber of Commerce supported it. Walking delegates of various unions enforced the boycott directly by standing in front of Chinese shops or businesses employing Chinese and advising patrons not to enter. After Chinese businessmen appealed unsuccessfully to the police for protection against such harassment, they filed a suit in equity seeking an injunction against such boycott activities. They were granted a temporary injunction in April 1897, and this was finally made permanent in 1900. But the boycott had taken its toll, especially among the poorer Chinese who needed steady employment to live and therefore had to leave Butte in order to survive.[30]

The Chinese population in the United States reached it peak in 1890 and declined steadily thereafter until 1950. A combination of the Exclusion Act and the boycotts had their effect in Anaconda and Butte. While newspapers played up incidents of smuggling Chinese across both the Mexican and Canadian borders and frauds with reentry permits, the Chinese population in Anaconda followed the national trend. The number of Chinese businesses which applied for licenses declined abruptly from 1893 to 1894, and the number of Chinese businesses listed in the city directory also decreased. By the turn of the century there were probably less than twenty Chinese in town. A report issued in 1902 indicates that by 1900, there were only 1,739 in the entire state. The campaigns against them subsided and those that were left became accepted members of the community, albeit still considered "peculiar".

But once again in 1902, when the exclusion period had run its ten year course, there was renewed activity to whip up public support for another extension. The Montana Bureau of Agriculture, Labor & Industry issued a report in 1902, which brought up all the old arguments for excluding the Chinese. In fact, much of it read as though it had been copied straight from the articles and editorials of the Deer Lodge *Montanian* of 1871. But new charges were also made and elaborated on, referring specifically to the rescue of two female Chinese slaves in Butte by a mission worker from San Francisco. They were held in slavery at 21 Galena Street (the red light district) and compelled to lead a life of shame. It went on to charge that Chinese prostitutes who contracted venereal diseases were confined to apartments reserved only for whites. It warned that even talking to a Chinaman might be dangerous since they suffered from a dangerous disease of the teeth known as "caries" which might be transmitted by their bad breath. And it cautioned against wrapping babies in clothes washed by Chinese, who may have leprosy. But it also recognized that attitudes were changing by admitting that "It is doubtful whether it is possible to arouse sufficient public sentiment against these foreigners..."[31]

Nevertheless, there was enough popular sentiment against them nationally to renew the exclusion act for ten more years in 1902 and once again in 1912. In 1924, the immigration law was made more restrictive across the board and resident Chinese were forbidden even to bring their spouses and other family members to join them.

By 1902, the boycott had already achieved its purposes in Anaconda. While there were still three or four laundries and a couple of noodle parlors, there was no longer a Chinatown. The color and variety of the earlier days were gone never to return. The newspaper stories about the Chinese were now limited to Butte and they reflected a decidedly romantic and more tolerant attitude. Now the emphasis was on such things as Chinese New Year's celebrations— full of color, frenzied hubbub, firecrackers and rice wine—and on Chinese funerals which were reported in rich detail, dwelling on the scattering of colored paper along the procession route to detain the evil spirits and the serving of a fine banquet at the grave to propitiate the gods. There was usually a mention toward the end of these articles about how white residents who observed these funerals seemed to find much humor in it all, and how local bums would finish off the food left on the grave as soon as the principals had departed.

Even in 1901, this change can be noted in a report on Fun Gee, a pioneer Dublin Gulch laundryman, who would send his men to the house of a poor white washerwoman with ten children to help her do her washing and ironing and would even distribute quarters and dimes to the children. It ended on this note: "Yes, Fun Gee is a Mongolian and a heathen, and they are not popular in the United States just now, but the honest man will be compelled to admit that if we had more citizens like Fun Gee this country would be a better place to live in."[32]

During the following years, the former sense of threat was replaced by a more friendly and even nostalgic mood. A feature story in the *Anaconda Standard* of May 20, 1906 conveys this, referring to the Chinese residents of Butte as peaceful and industrious citizens. "They care for their own people and don't burden welfare, they never bring a suit in court and are never sued...They are fast becoming Americanized and the mission in Chinatown is Christianizing many of them... Society people in Butte have made the Chinese noodle parlors popular places to eat...Chinese truck gardens on the flat provide the city with fresh produce...They measure up well to other foreign groups in town and are definitely more peaceful...While they cannot become citizens, they take it as it comes, neither whining or complaining." In a melancholy reminder, it noted, "Their numbers are getting smaller year after year."[33]

Even references to gambling and dope took on more tolerant tones. A police officer testifying at a trial on the operation of an opium den in Butte explained that it was difficult to actually catch anyone smoking because the building in question had a dozen or more trap-doors at ground level all opening on the alley. When a raid took place, the occupants of these cubicles would open the doors with a sharp kick, roll out into the alley and get away. There was no explanation of why the police couldn't cover the alley exits as well. In another case in Missoula, a Chinaman charged with selling liquor to Indians was let off because he didn't know it was against the law.[34]

Butte's Chinatown survived the boycotts and over the years has added much to the city's vibrancy and variety. In 1913, there was an article in the *Standard* about Yo Foo Kee, the first Chinese resident of Butte, who not only continued to live there, but had become a millionaire. Yo and others like him continued to make Butte a more interesting place through World War I and into the Thirties. By this time, the Chinese in Anaconda were only a memory.

11. A Small Company Town

The Montana constitutional convention of 1889 specified that the capital of the state would be determined by popular vote. On the first go around in 1892, Helena had come in first, Anaconda second, Butte third and Bozeman fourth. Since no city had a majority, a run-off between the two front runners was held in 1894. With Daly supporting Anaconda, it was only natural for W.A. Clark, who had pushed for Butte in the first round, to oppose Daly and throw his support to Helena.

It so happened that in 1893, Clark was making his third try at elective office, again for the U.S. Senate. Just as naturally, Daly was bound to support his opponent. As in the previous elections, money flowed freely from both sides. Since Senators were elected by the state legislatures, Daly needed more than his newspaper to achieve his objectives. Many legislators were selling their votes to the highest bidder and both political parties played the game.

It was the seat occupied by Col. Wilbur F. Sanders, that Clark aspired to fill. Sanders, a principled and unbending Republican was a hero of the vigilante days in Bannack and Virginia City, and had been active in bringing the Territory of Montana into existence. He had drawn the short three-year senate seat when the legislature named the states first two Senators in 1890. Unfortunately for him, the Democrats with the Populists held a majority in the joint session of the 1893 legislature. While Clark was able to secure the endorsement of the Democratic caucus, eight Daly Democrats bolted and proposed their own nominee, Daly's chief attorney, William Wirt Dixon.

This threw the joint session into a hopeless deadlock with none of the three candidates able to achieve a majority. After voting every day from January 10 to February 11, the Republicans attempted to gain support from the Daly forces by substituting for the venerable Sanders, the Republican mayor of Butte, Lee Mantel, a Daly ally. But the Daly forces would not yield and the legislature adjourned without naming a Senator. Republican governor, Lee Rickards, exercising his right to make an interim appointment, named Lee Mantel. However, the U.S. Senate opposed such appointments on the grounds that the legislature was required by law to elect a Senator, in a special session if necessary. They rejected Mantel by a narrow margin. Rickards refused to call a special session knowing a Democrat would win. So for more than a year Montana had only one Senator and W.A. Clark fumed over the cussedness of his rival Marcus Daly.[1]

This setback made Clark more determined than ever that Anaconda should not become the state capital and he opened his bank account wide to keep the prize away from Daly. Of course, Daly was not resting on past victories, either. He began increasing his economic activities all over western Montana, where the state's population was concentrated and where he was the largest investor. In the Bitter Root Valley and around Missoula, additional lumberjacks and sawmill operators were hired and immediately registered to vote. Matt Kelly says that three hundred men were taken in a special train from Anaconda to Butte, where they were marched to the ball park to join seven hundred more. There in front of a judge they raised their right hands and took the oath of allegiance to the United States and were declared American citizens. They then filed past the judge and his clerks and were given their citizenship papers.[2]

Daly's BA&P Railroad had begun regular service between Butte and Anaconda on January 1, 1894. With election drawing near, Daly renewed work on the road bed west of Anaconda. Full crews of graders were hired to prepare the way for new track. This, of course, was in the days when no public highways yet existed in Montana and the only overland routes besides the railroad were rough wagon roads or horse trails. Daly not only wanted to increase the numbers of workers loyal to him and his cause, he needed to demonstrate that his new capital, Anaconda, was not off the beaten path but directly linked to Missoula and the rest of northwestern Montana by a modern transportation system, the railroad.

He announced that the railroad would reach Granite, Philipsburg and Missoula before the election and that another branch would go to Hamilton and on to San Francisco. The *Anaconda Standard*, throwing caution (and veracity) to the winds, was predicting as early as October 23, 1892 that "The new railroad will be part of the best transcontinental system operating in the state. Its completion to the western boundary of the state within the next few months is a fixed fact—there is not a shadow of a doubt about it." Every day during the months before the vote, the *Standard* carried a half page ad which urged people to vote for Anaconda because it would have a railroad linking it to Missoula making it the best choice as a capital. But by election day, the roadbed extended only as far as Silver Lake, 16 miles west of Anaconda, and no rails had been laid.[3]

Daly recognized from the beginning that he was up against formidable odds. Helena could count not only on Clark's money but also on its own millionaire power broker Sam Hauser, and on support of the Northern Pacific railroad. In Missoula, Daly's former business partner A.B. Hammond abandoned him. And in Helena, Thomas Carter could not find it in his heart to support Daly against his own home town. While Daly got the immediate support of Jim Hill and his Great Northern railroad, he couldn't take for granted the vote of any town in the state except Anaconda itself and perhaps his other town, Hamilton, in Ravalli county. Even in Butte, where a majority of the miners worked for the Anaconda Company, loyalties were divided. Lee Mantel backed Anaconda, but Daly's former employers, the Walker brothers, favored Helena. So did the Boston and Montana and the Butte and Boston companies which campaigned vigorously for Helena. Each side worked hard to get newspaper support throughout the state. The *Great Falls Tribune* and *Bozeman Weekly Chronicle* editorialized in favor of Anaconda. But *The Missoulian* was outright nasty in its opposition, asking tartly: "What has Anaconda ever done for Missoula, anyway? If Christ came to Anaconda he would be compelled to eat, sleep, drink and pray with Marcus Daly,". Clark began buying up small-town newspapers around the state or providing them with subsidies to give favorable publicity to Helena.[4]

Frequent exchanges of verbal bombast between the *Standard* and the *Miner* were commonplace during the campaign. The *Miner* claimed that the Anaconda Mining Company was an octopus owning 8,149 acres surrounding Anaconda and that it owned 926 city lots, implying that the Company would also own the state government if it were located in Anaconda. The Clark forces manufactured and distributed thousands of copper collars as symbols of Anaconda Company control. Meanwhile, the *Standard* always referred to Helena as the "temporary capital" and charged that Helena was importing voters from out of state, that it was a racist community, discriminating against negroes and was too high-brow. At that time the number of blacks in Anaconda was less than twenty.[5]

To reenforce the image of Helena as a foppish, elitist city, Charles Eggleston, Assistant Editor of the *Standard* and an accomplished satirist, wrote a pamphlet which he attributed to Helena boosters purporting to ridicule Anaconda and its pretensions:

"A rude, rough smelter town rooted in vileness and vulgarity; a town nine-tenths of whose population toil the year round on manual labor; big, strong, coarse working men who could not tell a German from a wheelbarrow; that so far from exhibiting a sense of mortification and chagrin; they seem to take a sort of unconscious pride in going to and from their work in soiled overalls and with huge dinner buckets; laborers, mechanics, artisans, bricklayers, copper-dippers, whose average wage reach only $105 a month...

"Spending their time cultivating their little sawed-off gardens, going to low picnics and organizing and perfecting labor unions, which, Helena is informed and believes, are the greatest curse of the modern world.

"The whistles that blow every morning at seven o'clock, so soon after society people have retired, are an intolerable nuisance. It is impossible to walk the streets of Anaconda without seeing working men and their wives and children, and when the streets are crowded one cannot escape brushing up against them."

A Helena pamphlet questioned the wisdom "of the removal of the seat of government to a town located about fifty miles from the southwestern boundary of the state and owned by a rapacious corporation."[6] Anaconda was cast as a small company town off the beaten path, connected only to Butte by a little railroad that ended in a gulch a few miles west of town. Much of Helena's vituperation was concentrated on James Ben Ali Haggin, of Turkish descent on his mother's side, rather than on Daly, who remained personally popular. Haggin was caricatured as a Turkish merchant wearing a fez, "Mr. Haggin, of New York, San Francisco, Deadwood and Constantinople."[7]

Harpers Weekly ran an article about the fight, written by a former editor of the *Helena Journal,* which called Anaconda an isolated out of the way place in an obscure corner of Deer Lodge Valley. The Northern Pacific transported over 10,000 people to

Helena from all over the state free of charge "to visit the city and get to know it". Even religion became an issue with the anti-Catholic American Protective Association disparaging Anaconda in its newspaper as being too Irish and too Catholic.[8]

This was an especially touchy subject among the Butte and Anaconda Irish. Butte had become a western haven for poor, jobless Irish wanting to start a new life. Unlike many cities in the East where signs were posted at job sites, "No Irish Need Apply", Daly and his mine foremen gave preference to immigrant Irish laborers looking for jobs as muckers in the mines. There was an active and sometimes bloody rivalry in Butte between Catholic Irish and Protestant English miners from Cornwall, who the Irish derisively called Cousin Jacks. The Irish worked mostly in Anaconda mines, the Cornish in Montana & Boston, W.A. Clark mines or other mines on the hill. The Irish were in the majority, dominating the Miners Union, Butte political offices, stores, and saloons, as well as many supervisory positions in the Anaconda Company. Butte was their town. Like the Mormons who had founded a western citadel in Salt Lake City against a hostile eastern establishment, Butte was a western refuge for the Irish.

While Anaconda had a large Irish Catholic population, there was greater balance here among ethnic groups in terms of numbers, and their rivalries much more subdued than in Butte. If the Anaconda smelter showed any preference for hiring the Irish, this was not evident from its polyglot payroll lists or the ethnic mix in the town itself. There were large numbers of Scandinavians, and already southern European names began to appear on various local business houses. But Anaconda was not exempt from the winds of the Protestant Anglo-Saxon crusade which blew across the U.S. in the late 1880s and 1890s that touted British racial superiority. Nativism and anti-Catholicism inspired organizations like the American Protective Association. Its charge that Anaconda was too Irish and too Catholic to be the capital was a barb that stung the Anaconda Irish.

On election day, both sides had barrels of whiskey on the streets of Butte giving away drinks and handing out five dollar gold pieces at the polling booths. Daly admitted to spending $350,000 on the campaign. Clark accused him of spending at least $1 million. Through these campaigns bribery and corruption were becoming an accepted part of political life in the young state.[9]

In the early hours of the vote count, Anaconda held a slight lead, 16,853 against 16,266 for Helena. The vote from Missoula was disappointing to Anacondans because it broke evenly, with Anaconda gaining by only 220 votes. Daly supporters had expected better from the lumber interests. The final count showed that Helena had won with 27,028 votes to Anaconda's 25,118. Matt Kelly says that the tailors union of Butte was responsible for Anaconda's defeat, without saying why. The Butte vote was only 6,513 for Anaconda and almost as many, 4,003 for Helena.[10]

But there were some other interesting anomalies in the vote. Ninety-five per cent of the voters in Helena's Lewis and Clark county voted for Helena, while only eighty-one per cent of the voters of Deer Lodge County voted for Anaconda. There were 995 residents of Deer Lodge County who voted for Helena. Had they all voted for Anaconda, thus taking away that many from Helena, the final count would have been Anaconda 26,073 to Helena 26,069 and Anaconda would have won. Most of the Deer Lodge County votes for Helena were cast by residents of Deer Lodge and by those in the far northern corner of the county around Avon, Elliston and Helmville, where 14 precincts voted a majority for Helena. Not only were they closer to Helena than to Anaconda, there was some resentment that upstart Anaconda had beat out Deer Lodge, one of the oldest and best established towns in the state, in the first round of balloting. Also, Deer Lodge was the original headquarters for W.A. Clark's Bank and he still exercised influence in the Community. Interestingly, the largest proportion of votes for Anaconda came not from Deer Lodge County, but from Daly's other town, Hamilton in Ravalli County.

Location undoubtedly played a role in the final choice. Helena was more centrally located. All of the eastern counties voted unanimously for Helena. Also, there is no doubt that many voters heeded Clark's propaganda and rejected putting their state capital in Daly's company town, thus symbolically removing it from his influence. But whether in Anaconda or Helena, copper was still king. It had come to dominate the life of Montana. As if to advertise this fact, when the state capitol was built, its dome was sheathed in copper.

After two painful defeats at the hands of Daly, it was now Clark's turn to crow. He had deprived his hated rival of the thing he had wanted most. What a sweet victory it was. When he arrived in

Helena by train on election night, ecstatic crowds unhitched the horses from his waiting carriage and pulled it through the streets themselves. Clark invited the whole town to drinks on him, and the saloons were soon emptied of their entire stock. It is reported that his bar bill for the evening was $30,000, but this was only a small fraction of what he'd spent to achieve his aims. And to him it was worth every penny. He had not only beaten Daly, but he'd gained an appreciative and loyal following. He was Helena's hero.[11]

Because of the charges made against the Irish during the campaign, Anaconda's defeat was taken as an affront to their national pride by the Irish societies of Butte and Anaconda. Thus soon after, a campaign was mounted to erect a statue on the capital grounds in Helena to an Irish hero, Thomas Francis Meagher, who had been acting territorial governor and who mysteriously drowned while on a steam boat near Fort Benton. Marcus Daly gave the campaign his full support, matching contributions from other sources. Not long after the capitol building was completed, the statue was placed at a prominent spot in front of the capitol and dedicated by Catholic Bishop Brondel, with large delegations from Butte and Anaconda in attendance.

The selection of the state capital appeared to mark a turning point in Daly's interest in Anaconda. While he continued to invest in and expand his smelter operation, he no longer lavished the same attention on the town that he had previously. His other town, Hamilton, in the Bitter Root Valley, now became the apple of his eye. He spent whatever leisure time he had at his new mansion there concentrating on his ranch and stables of thoroughbred race horses.

There were business reasons, as well, for him to spend more time away from Anaconda. Supplying timber for the mines was a growing concern for him. They were by this time consuming from 40 to 50 million board feet a year. Having used political leverage to get the suits against the Montana Improvement Company dismissed, Daly increased his holdings around Hamilton, buying six million board feet of standing timber in northwestern Montana and negotiating with the Northern Pacific Railroad for even more. By 1898 he had bought 700,000 acres with an estimated stand of more than 200,000 feet per acre, more than 140 billion board feet of timber.[12]

He established saw mills at Hamilton, at St. Regis, further north, and at Hope, Idaho. He purchased the Blackfoot Milling

Company from R.A. Eddy and A.B. Hammond, which included a
large saw mill near Bonner and 90,000 acres of timberland, and
bought out the Montana Lumber and Manufacturing Company he had
earlier developed with A.M. Holter. He also formed the Blackfoot
Land and Development Company to convert logged-off land into
farms. At about the same time he purchased coal mines at Belt near
Great Falls, at Storrs and Cokedale near Bozeman Pass, at Washoe
and Carbonada near Red Lodge and at Diamondville, Wyoming to
supply fuel for his new smelters.[13]

During this time, Daly had to appear in court concerning
some pending litigation against his company. When he was sworn in
he was asked to give his name and his permanent residence. He
answered that he was constantly traveling and divided his time
between residences in Butte, Anaconda and Hamilton. The clerk of
the court then said, "For the record, sir, where do you spend most of
your time?" Daly smiled and said, "At Garrison, waiting for the
train." Garrison was the connecting station for the Northern Pacific
Railroad, which took Daly to Missoula and thence by horse carriage
to Hamilton.[14]

The town of Anaconda was, thus, left to its own devices.
Talk before the elections about Anaconda becoming the home of the
new University of Montana if it conceded the capital to Helena was
heard no more. In fact, the legislature in 1893 had designated
Missoula for the University, Bozeman for the Agricultural College
and Butte for the School of Mines. Helena natives, no doubt,
considered it entirely fitting when Anaconda was awarded the new
state insane asylum established at Warm Springs a few years later.

In the midst of the capital campaign Anaconda was fighting
a serious smallpox epidemic which broke out in the spring of 1893
and raged into early 1894. The primitive medical knowledge of the
time was invoked and a pest house for women patients attended by
nuns from the hospital was opened nearby. The hospital itself was
quarantined for three weeks during which no one was allowed to
enter or leave. Communication with the hospital was maintained by
a rope attached to the doorbell. Men patients were removed to a log
cabin built in an isolated gulch southeast of town. Here they were
attended by other men who had had smallpox previously. Most of
those stricken with the disease died and were buried in a quarantined
graveyard further up the gulch.[15]

Those buried there were soon forgotten, but for the next fifty years or so, the gulch and its graveyard were known as Smallpox, and in the 1930s during the winter, young boys from the town's new eastern addition, innocently used the gravestones to build ski jumps.

12. A Uniform Wage Underground

The reckless exploitation of natural resources and the native peoples in the West was, of course, a reflection of the times which expressed itelf in many ways including the exploitation of wage earners. As late as 1900, seventy percent of industrial workers worked ten-hour days, seven days a week. In the steel industry, twelve-hour days were the norm until 1923. The average work week in textiles was 60 to 80 hours. Trainmen put in 70 hours. New York City bakers worked 80 to 100 hours a week. In 1891, a railroad laborer earned $1.75 a day. In 1895, a common laborer working on the reservoir in Croton, N. Y. earned a maximum of $1.50. From 1880 to 1910, the average income for unskilled workers was less than $10 a week; and even for skilled workers, rarely more than $20. During this entire period the average annual income of industrial workers was never more than $600. This, combined with periods of high unemployment, meant that large segments of the population were living in misery if not abject poverty.[1]

Under the circumstances, it is understandable why people put up with the hardships of frontier life to improve their lot and why the lure of gold was so strong to so many. Even for those who didn't strike it rich, the mines of the West offered a better income than most work in the industrial East. By the turn of the century Butte was known as one of the highest wage towns in the country. Of course, better wages didn't come without confrontation and conflict.

As early as 1863, miners working in Virginia City, Nevada had organized themselves into a "Miners' Protective Association" to maintain the daily wage for all underground work at $4 a day. Then because of hard times and black-listing of union members by mine owners, it was reduced to $3.50. However, in 1867, a new Miners Union was able to reinstitute the $4 daily wage. This was justified on the basis of high earnings by the mine owners and the high cost of living in a mining camp. Before long, all of the trades connected with mining in the Comstock region had their own unions. Remarkably, the Miners Union was able to establish an eight hour day for underground work in 1872.[2]

Since many of the miners that made their way to Butte in the late 1870s came by way of Nevada, it is not surprising that they brought with them a strong belief in united labor action. Thus when the Butte mines became larger as investments increased and the labor force grew, the miners formed unions to bargain with management over wages and hours. Marcus Daly, who had been a miner himself in Virginia City and had benefited from union bargaining, cannot have been surprised or unsympathetic when he discovered union activity at the Walker brother's Alice Mine which he managed. Nevertheless, in 1878, Daly, on advice from Walker brothers representative William Read, and in agreement with A.J. Davis, the manager of the Lexington mine, cut the wages of unskilled miners from $3.50 to $3.00 a day. The miners, ably led by Aaron Witter, went out on strike, and then successfully bargained with Daly to establish a uniform rate of $3.50 a day for all underground workers. This was a breakthrough for the union because generally mine owners insisted on a differential between skilled workers, who were relatively few in number and unskilled, who were many. It was considered such an extraordinary triumph that the anniversary of the victory, June 13, became a local holiday in Butte, with parades, speeches and a grand labor ball. From that time until internal conflict eventually brought the union to its knees in 1914, June 13 was celebrated as Miners' Day in Butte.[3]

There was more at stake in the resolution of this conflict than was apparent at the time. It introduced Daly's philosophy and management technique which contrasted with commonly accepted managerial practice of resisting all the miners' demands, locking them out if necessary and hiring new crews, counting on the traditional wandering spirit of the miners and the free-wheeling atmosphere of the mining camps to replenish the work force. This was the well-established pattern throughout the west and contributed to the instability and transient nature of most mining towns. Daly's decision, which was resisted by Read, laid the groundwork for two very important developments, the initiation of a period of labor peace in Butte which lasted until after Daly's death, and the establishment of a core community of permanent, skilled labor for the mines. These two factors would change the labor picture in Butte from that of just another western mining camp into one more akin to a growing industrial center.[4]

The miners' victory had other far reaching effects because it established Butte as a center of successful union activity. This could not have been good news for the Walker brothers, but Daly took it in stride. The Butte Miners' Union grew steadily and, when it inaugurated its large multipurpose hall in 1885, it had 1,800 dues-paying members. The other trades and crafts connected with work in the mines soon formed their own smaller craft unions, which, along with the Butte Miners Union, established the powerful Silver Bow Trades and Labor Assembly in 1886. By 1900 there were thirty-four separate unions representing most of the 8,000 workers in Butte. It became the largest and strongest labor association in the West, referred to later as the Gibraltar of unionism.[5]

The Butte Miners Union was one of the earliest affiliates of a national labor movement which graced itself with the sonorous name of the Noble Brotherhood of the Knights of Labor. The Knights agitated not only to boycott Chinese labor (as discussed earlier), but for an eight-hour day, for the abolition of child labor, a national tax on income and for other social and economic reforms. In 1884, The Knights of Labor used its organizing skills to help rail unions win a major victory in the Southwest. This gave the fledgling labor movement a shot in the arm and sent shudders through management offices across the country.

In 1886, with wages falling and discontent rising, the Union Pacific and other roads had continual problems with the Knights. In March, a strike which started on the UP in Kansas City spread to the Utah and Northern and turned ugly. At Butte, where "the hardest people to be found anywhere were congregated", cars were derailed, air hoses cut and valves smashed. The Wyoming and Idaho divisions of the UP, of which the spur to Butte was a part, had a reputation for having the toughest crews on the line to deal with.[6]

Several months later, the Knight's campaign for an eight hour day suffered a serious setback in the Hay Market bomb incident of 1886, in Chicago, where seven people were killed and over sixty injured. Even though the Knights were in no way responsible, their influence on the labor movement was never the same after that. Interestingly enough, in that year two cigar makers, Samuel Gompers and Adolph Strasser, founded another union: The American Federation of Labor (AFL).[7]

Anaconda, Butte's loyal sister, was from the time of its birth a town with strong union leanings. The counterpart of the Miners' Union in Butte, and the largest in Anaconda, was the Mill and Smeltermans' Union. In addition, each of the trades had separate unions. In his *Comical History of Montana*, Jerre C. Murphy says that "Butte was the strongest union town on earth", a place where no employment was possible for a man who did not belong to a union. There was even a chimney sweeps union, composed of two chimney sweeps.[8]

Anaconda was but an extension of Butte in this regard with separate unions for Barbers, Building Laborers, Bartenders, Bricklayers, Brickmakers, Blacksmiths, Brewers, Butchers, Carpenters, Cigarmakers, Clerks, Construction Workers, Cooks, Iron Molders, Locomotive Firemen, Machinists, Musicians, Painters Decorators and Paperhangers, Shoemakers, Switchmen, Teamsters and Typesetters. There was even the Anaconda Master Horseshoers' National Protective Association, Local 184. Lodge 29 of the International Association of Machinists, organized in Anaconda in 1888, was the first local unit of that organization west of the Mississippi. In the Anaconda city directory of 1899 the Deer Lodge Trades and Labor Council lists 2,500 members. While Daly as owner and manager of the Anaconda reduction works could have resisted this development, he understood his workers and never became anti-union.[9]

In fact, during these years his *Anaconda Standard* had a strong printer's union and its editorial policy was supportive of the unions and gave their activities prominent news coverage. A report of Labor Day in Butte in 1894 is typical. Under a headline of "Solidarity of Labor in the Two Cities", a front page story went on to describe how between 1,500 and 2,000 visitors from Anaconda joined with their brothers of organized labor in Butte to celebrate labor day. Over 1,000 of them marched in the parade, accompanied by the Copper City Band of Anaconda. The first division of the parade was composed of fourteen unions from Anaconda headed by James H. Hoy. The Blacksmiths union, performing a drill with big hammers as they marched, won the prize of $50 for best appearing organization in the parade.

Delegates from the Silver Bow Trade and Labor Assembly marched from their hall on north Main Street to meet and welcome at the BA&P Depot the Anaconda Unions which arrived on a special train of 19 coaches with over 2,000 passengers. "The fourteen Anaconda Unions with their magnificent banners formed in line on Utah Avenue headed by the Copper City Band and marched uptown behind the delegation of the Silver Bow Assembly. Trains carried passengers free of charge all day long..."[10]

This picture of labor peace did not prevail in other mining centers, however. The issues in the vast majority of miners' strikes across the West were consistently the same, the eight-hour day and the differential for underground work. In some cases the unions sought to set the wage at $4.00 in others at $3.50. Just as consistently the owners opposed these demands and were generally able to utilize the power of government to bolster their resistance. In 1880, a miners' strike in Leadville, Colorado over hours and wages was crushed by the state militia. Such government intervention was to become the pattern in most mine, mill and smelter labor disputes throughout the West.

The eclipse of the Knights in no way diminished the activism and crusading spirit of the Butte Miners Union, which sent organizers into Idaho and helped found locals in Wallace and Mullen on the Coeur d'Alene River and in nearby Gem and Burke on Canyon Creek and was able to establish the Butte uniform wage rate of $3.50 for a ten hour day for all underground work. When owners tried to reduce the rate to $3.00 in 1891 for unskilled workers, a brief strike reestablished the Butte rate.[11]

The struggle was not over, however, and to its credit, Daly's *Anaconda Standard* published regular reports on happenings in the Coeur d'Alene, since many in Butte and Anaconda had friends and family there. The next year the Union Pacific and the Northern Pacific railroads raised their rates on ore shipments and the Idaho mines shut down leaving 1,600 unemployed miners to shiver through the winter. After three months the railroads reestablished the old rates and the mines reopened on April 1, but with a differential for unskilled workers of $3.00.

The unions called another strike and were supported in their action by a relief fund of $30,000 sent by the Butte and Anaconda unions. The owners brought in scab labor and the strike dragged on into July. On July 9, a Butte organizer recognized one of the local union officers as a Pinkerton spy and all hell broke loose.

The union members attacked the scabs, and security guards fired on the union men. The striking miners blew up the mill of the Frisco mine in Gem, killed two security guards and ran all of the scabs out of the mines along Canyon Creek. The Governor called in the Idaho National Guard, supported by 1500 U.S. troops. Three hundred and fifty miners were rounded up and put into a "bullpen", which consisted of old boxcars and empty warehouses surrounded by a fourteen foot stockade fence. The strike and the union were broken. The *Anaconda Standard* editorialized that the mine owners "will prosecute and persecute with fiendish vengence" and argued for a general amnesty. In Butte, Marcus Daly contributed funds to help families of Butte and Anaconda union men imprisoned in Idaho.[12]

After the union men were freed, a meeting was held in Butte on May 15, 1893, to take stock of the setback in Idaho. Representatives came from mining unions in South Dakota, Utah, Idaho, Colorado and Montana. Out of that meeting came a new organization, the Western Federation of Miners and the Butte Miners Union became Local 1 of the new federation. In an editorial on May 21, the *Standard* endorsed the formation of the federation but took exception to some of its objectives such as government ownership of all railroads.[13]

The new WFM would soon have plenty to contend with. Their meeting coincided with the start of the panic of 1893. But the nations economic woes had been preceded by an even more deep-seated political stagnation. It was evident by 1890 that neither the Republican nor the Democratic party reflected the views or addressed the needs of the nations farmers or of its growing numbers of industrial workers. This resulted in a political revolt and the formation of the national Populist party in 1891. At a convention in Anaconda the following year the Montana Populist party came into existence.

13. Knowledge and Culture

On the national scene, the failure of the National Cordage Company, quickly followed by that of the Reading Railroad brought on the panic of 1893. By July of that year, the Erie Railroad went the same route, followed by the Northern Pacific, the Union Pacific and the Santa Fe. Within two years one-fourth of the railroad capitalization of the country was under control of bankruptcy courts and sixty percent of railroad stocks had suspended dividend payments. Before the year was out, over fifteen thousand businesses went under and 158 banks failed, 153 of which were in the South and the West. In Denver alone, twelve banks closed their doors for good. The demonetization of silver the same year caused the price of silver to collapse from 87 cents an ounce to 62 cents. Silver mines shut down throughout the West; and by the summer of 1894, four million jobless walked the streets of factory towns in search of work.[1]

This panic and the resulting depression scarcely touched Anaconda. It was an island of industrial activity in a sea of economic troubles. Following the repeal of the Sherman Silver Purchase Act, Granite, the silver town to Anaconda's west, closed down completely, as did the Alice, the Lexington, the Blue Bird and smaller silver operations in Butte. One third of Montana's work force was unemployed. Banks, including Sam Hauser's in Helena, failed. Daly wrote his partner Haggin, "Butte is looking very savage, there are over 3000 idle men on the streets. They are discontented and dissatisfied." He opposed suggestions that wages be cut, arguing that this would destroy workers' families and that it could lead to bloodshed and property destruction.[2]

While copper prices were depressed, demand continued to grow and production from both Butte and Anaconda increased. Daly kept the Anaconda work force at full strength and continued modernizing his facilities. The converter plant, installed in 1892, with twelve converters producing 225 tons of blister copper every twenty-four hours, maintained full production. The electrolytic refinery, which opened the same year, was expanded so that more and more of the smelter's blister copper could be refined locally prior to shipment and sale.

Capital or not, Anaconda continued to grow and attract new residents. It retained the fighting spirit instilled by Daly and it had a momentum built up by ten years of uninterrupted investment and growth. New construction in the town continued throughout the 1890s and many of the landmark buildings, which would define the city center for the next eighty years, were erected at this time. The town was taking on that appearance of substance and permanence which it would preserve during the entire twentieth century. Anaconda, like the country itself, had a confidence and an optimism that things could only get bigger and better.

The Hoge, Daly and Company bank, renamed the Anaconda National Bank, moved one block from its one-story building on Main and Commercial to new headquarters in a striking new two-story brick building on Main Street diagonally across from the Montana Hotel. The *Anaconda Standard* had a home of its own in a large, handsome, two-story brick building on the corner of Third and Main streets, built to John Durston's specifications. The owners of a cigar factory, Joseph Steigler, Joe Paul and John Petritz built a new brewery on the northwest corner of Walnut and Fourth Street, replacing R. Fenner's brewery which had burned to the ground in 1888. The Ancient Order of Hibernians had finished its new building and the city hall across the street was a three-story gothic stone and brick structure, complete with an imposing clock tower and large enough to house the police department, city jail, fire department and municipal offices.

By 1899, the central section of the town had forty brick buildings, each of two or three stories, known as blocks, buildings or halls. Some of the better known and longest lasting were the Copper City Commercial Block, the Durston Block, the Flood Block, the Parrot Block, the Petritz Block, the Monitor Block, the Beaudry Building, the Barich Building, the Davidson Building, the Dwyer Building, the Fortier Building, the Sheilds Building, the Stagg building, Dewey Hall, Evans Hall, Salvation Army Hall, Silver Hall and Turner Hall.[3]

In 1897, construction began on two other buildings on Main Street, less than a block apart, which would not only add distinction and class to the city's architecture, but would contribute to its refinement and become the center of its cultural life. Not surprisingly, two women were connected with these endeavors, Margaret Evans Daly and Phoebe Apperson Hearst.

Phoebe Hearst visited Anaconda for the first time with her husband, George Hearst, in 1885, and took an immediate interest in improving its cultural life. Mary Dolan, in the 1983 collection of her newspaper articles on Anaconda, reports that, after receiving no suggestions from the city fathers when she asked them what she could do for the town, Mrs. Hearst proposed that she found a library. She was then told that, while this was exactly what the town wanted and needed, they hadn't asked because they realized it would cost a great deal of money and they didn't want to impose on her generosity. This reply endeared the town and its citizens to her and sparked a relationship that lasted until her death in 1919.[4]

George Hearst, who was already active in the California Democratic party, took over the faltering *San Francisco Examiner* newspaper in 1880, when its owners could not repay loans he had made to them. In 1887 he was elected by the California legislature to the U.S. Senate, and turned over the *Examiner* to his son, William Randolph Hearst, who had been thrown out of Harvard College several years earlier. Phoebe Hearst, who was a former school teacher and much younger than her husband, moved with him to Washington and became active in promoting educational endeavors. She founded the National Cathedral School for Girls there and made financial contributions to numerous other educational projects in California and elsewhere in the U.S.

When Senator Hearst died in 1891, he left his entire estate, including the *San Francisco Examiner* and his interest in the Anaconda Copper Mining Company to his wife. In 1895, she sold her one-quarter interest in the Anaconda Company to the Rothschilds of London for $7,500,000, and gave the money to her son, William Randolph, to help him with his newspaper business. This sale set in motion two very different chains of events. It was the first step in transferring control of the Anaconda Company from San Francisco to Wall Street, and it was the beginning of the Hearst publication empire. Hearst immediately bought the *New York Morning Journal*, which was the first newspaper he owned outright, since his mother still owned the *Examiner*.[5]

Perhaps it was this sale of the Anaconda stock which reminded Mrs. Hearst of her promise to the city fathers of Anaconda, because that same year she opened a reading room at 308 Cherry street, stocked with 2,175 volumes and the leading periodicals of the time. This proved so popular that soon after she decided to construct a new library building in memory of her husband. She hired architect, S. Van Trees, of San Francisco to design it and T.C. Twohy of Anaconda to supervise the construction. Ground was broken on July 28, 1897 and the new building was inaugurated eleven months later, on June 11, 1898. That same day Mrs Hearst was made an honorary member of the Anaconda Women's Literary Club and it was announced that the club would make its home in the new building.[6]

The two-story structure, which stands on a large lot on the corner of Fourth and Main streets surrounded by well-maintained lawn and shrubbery, radiates the same splendid elegance today as it did on the day it opened. Built of locally made brick and Gregson granite in neo-classical style, its facade is dominated by two massive granite columns topped with Corinthian capitals. The great spreading arches of its first floor windows create a harmony for the entire structure that gives it a sober but distinguished grace befitting its function. The library entrance of two large oak doors is approached up eight granite steps and through a deep portico. Above the doorway on an arched marine tablet is inscribed "Hearst Free Library". Against the south wall of the main reading room on the first floor is a great fireplace with a large oak mantel reaching two-thirds of the way to the ceiling, and on the mantel, a bust of Senator George Hearst. From the day of its opening, it ranked with the Montana Hotel as the most elegant building in the city.[7]

After the building was completed, Mrs. Hearst continued to maintain the library and provide it with new books at an estimated cost of over $7000 annually, until 1903. In that year, Matt Kelly says, when a Socialist county assessor was elected who attempted to tax the library, Mrs. Hearst decided it was time to transfer ownership to the city. Always gracious, she also pledged $1000 a year for three more years to buy new books.[8]

William H. Thornton, one of Daly's partners who served as treasurer in both the bank and the hotel, became mayor of Anaconda in 1893. He held this office until 1897. In 1895, Thornton along with J.H. Durston, E.G. Smith, A.P. Cloutier, W.W. Reynolds, A.M. Scott,

E.S. Maxwell and M.J. Fitzpatrick organized the Anaconda Theatre Association to build an opera house in Anaconda.[9]

Up until that time whenever traveling theatre companies came to Anaconda they played at the various halls in town, which doubled as theaters, the Evans and Turner halls being the best known. Before these there had been the Cohen and Daly halls and the City Auditorium. But all were considered too small and inadequate for the growing and prosperous new town. The Evans Hall was housed in building on Main street with outside stairs leading up to its second-floor entrance. Also known as the Evans Opera House, it had become the most active theater in town for traveling companies. Chairs, set up for theatrical presentations were removed for dances and other events.

A humorous story about the limited facilities at these early theaters is attributed to John Maguire, a well-known Irish minstrel player who owned and managed the opera house in Butte and managed Evans Hall and later the Margaret theater in Anaconda. Madame Janauschek, a well-known star and singer of the day, was performing on the makeshift stage of the City Auditorium delivering an impassioned speech when she stepped on a loose plank. Being a very heavy woman, the plank went down under her and its other end shot into the air. The audience was at first startled then became convulsed with laughter. Janauschek vowed never to play Anaconda again.[10]

A similar story from the reminiscences of another Anaconda resident from those years also testifies to the lack of facilities. Otto Kloepful who lived in Anaconda in 1887, wrote to J.P. Braus, Secretary of the Anaconda Musicians' union, that when " 'The Devil's Auction' was put on in the old rink...a crew of men had to brace the platform with their backs between the [wooden]horses so the 40 girls could dance."[11]

Thornton and his associates believed it was time that the city, which led the state with its luxury hotel and its first-rate newspaper, deserved an opera house of equal distinction. They raised $15,000 through private subscriptions and enlisted the help of Margaret Daly to raise additional funds and interest her husband in the enterprise. The theater would be named "The Margaret" in her honor. At a public meeting in April 1896, chaired by Thornton, arrangements for the public subscription of funds were discussed and plans for the construction of the theater were announced. For building the theater, Marcus Daly pledged through his Townsite Company to donate the land and to match the total public subscription. Construction bids were received from some of the most prominent and qualified theater construction companies in the country: Henry Carter of Minneapolis, Renvic, Aspen, Wall and Renwick of New York; Wool and Lovell of Chicago and Frank Knox of New Orleans. Henry Carter got the job.

The building was finished in 1897, and was inaugurated on September 28 of that year. It was constructed at a total cost of $60,000, with the Townsite Company owing $30,000 of that amount. The play bill for opening night bore the following inscription: "Grand Gala Night. Dedication to the drama of the handsomest, largest and best-appointed edifice in Montana, the Margaret Theatre, John Maguire, manager."

This three-story building lacked the exterior grace and elegance of the new Hearst Library still under construction just up the street, but it was large, functional and luxurious on the inside. Its three-door entrance opened onto a large lobby, a ticket office, rest rooms, and the manager's office. Through swinging doors off the lobby was the foyer which led to broad, carpeted promenades and plush boxes along the walls.[12]

Anaconda and Butte were considered good show towns and were easily accessible to traveling companics because three major railroads came into Butte. For some of the major companies travelling to the West Coast, Anaconda and Butte were their only stops west of the twin cities.

The Evans Opera House was particularly busy, with a change of programs two or three times a week. A typical schedule (for March 1890) started with the Andrews Dramatic Company on a three day stay, presenting a different play every night, Pygmalion and Galatea, Uncle Josh and Rip Van Winkle, "popular prices 25 cents, 50 cents and 75 cents. Tickets on Sale at Playters." Then came "Maggie Mitchell, America's favorite actress, admission $1.00, gallery 50 cents," for two nights, followed by Ray Franchon in "Bunch of Keys, the funniest comedy on the American stage".[13]

Whenever it wasn't booked for traveling companies, Evans Opera House became Evans Hall and was solidly scheduled with dances and balls by the Ladies Guild, (tickets on sale at Playters) the Switchman's Mutual Aid Association, The Wolf Tone Guards, the

Fire Departments of the Upper and Lower Works and innumerable other organization. In fact, the hall got such heavy use that it was closed in the summer of 1891 so the gallery floor could be raised "since it was sagging a bit".[14]

Otto Kloepfel provides a colorful description of goings-on at the Evans Opera House from an even earlier time: "Professor Kennecot was the music teacher of the town. A four-to six-piece orchestra, led by Professor Wolf, used to play, "O Fair Dove! O Fond Dove" potpourri by Schiepegrell at the shows in Evans opera house. I was one of the audience that waited there until 1 a.m. for the snowbound Faust troupe to get in from Stuart, where Otto Wommiesdorf used to run a saloon. When the show was ready to begin Mephistopheles in his red devil suit stepped before the curtain and thanked us for waiting; and everybody stuck until the finish at 4 a.m."[15]

It was at Evans Opera House that Mark Twain, regaled a full house with his humor on August 2, 1895. John Philip Sousa and "His Unrivaled Band" played at Evans Opera House March 6, 1897. Tickets sold for $2.50 and $1.00. Most of the boxing heavy-weight champions appearing in their own shows, played Anaconda. These included John L. Sullivan, Bob Fitzsimmons, Gentleman Jim Corbett and Jim Jeffries.[16]

Anaconda was already on the vaudeville circuit of several well-known companies before the Margaret was built, and many independent road shows, chautauquas, noted artists and troupers played local houses regularly. The new Margaret made Anaconda an even more attractive venue for visiting performers. Of course, avid theater goers always had the option of catching one of the four daily trains to Butte for those special occasions when an internationally renowned artist was performing there.

In February 1891, Maguire's Butte Opera House presented four delightful nights of opera courtesy of the Emma Juch Grand English Opera Company, with "Tannhauser", "Carmen", "Les Huguenots" and "Il Trovatore". Sarah Bernhardt, the celebrated French actress came to Butte in 1906 and performed "Camille" one night and "Theodora" the next in French at the Holland Street Roller Rink. There were strongly differing opinions about the success of these engagements. The *Anaconda Standard,* commenting on "Camille" was decidedly negative, complaining that no one understood what was going on, and that even if the play had been in English the

Holland Street Rink was not a theater and the acoustics were terrible. It had a capacity for 5,000 and an audience of only 1,200. John Maguire, the impresario who had brought her, had a completely different story. He claimed that Sarah Bernhardt left Butte convinced that it was not only one of the friendliest towns in the United States, but one of the most cultured. Maguire said that after her performance in "Theodora", she was invited by Senator Clark, who spoke some French and whose young second wife was educated in France, to an elegant dinner in his house joined by some French speaking Butte residents, where she thoroughly enjoyed herself. He said that there were many French Canadians in the audience that evening who understood French, knew the play, and knew when to applaud and when to laugh. "She left Butte thinking everyone in town spoke French," he boasted.[17]

14. Free Silver and War With Spain

A principal reason for widespread discontent across the country grew out of a national policy of deliberate and long-term deflation, which made the value of money rise while the prices of goods and services dropped. The Civil War was largely financed in the North by printing paper money. Since this caused inflation and dollar devaluation, post-war administrations felt it necessary to strengthen the dollar. Industrial leaders adapted to this situation by holding wages down and introducing labor-saving devices and mass-production techniques. They also formed "pools" to keep the prices of their products high enough to make a good profit. The pools limited competition and fixed prices. Within a few years they were superseded by a new, more efficient mechanism, the "trust". A trust was a holding company which acted for all the companies which formed it. The first, the largest and the most famous of the trusts was John D. Rockefeller's Standard Oil Company. By 1880 Rockefeller controlled through his trust mechanism 90% of all U.S. petroleum production, which in the days before the automobile was used primarily to provide kerosene for lamps.[1]

This impulse to rationalize, integrate and consolidate large-scale production was the guiding principle of American industry during this period. J. P. Morgan and his United States Steel Corporation, the American Tobacco Company, the United States Sugar Refining Company, the Pullman Palace Car Co., etc. all followed in Rockefeller's footsteps. The trusts not only had enormous political influence but exercised absolute economic control over large portions of the United States. Whole states were under the sway of one group or another. In Wyoming it was the Rockefellers; in Utah, the Guggenheims; in Colorado, the Rockefellers *and* the Guggenheims; in Idaho, Bunker Hill and Sullivan, and in Oregon, the Union Pacific and the lumber companies. While Haggin and Daly, with their Anaconda Company were not yet in this league, they were moving fast in that direction.[2]

Unlike the large industrialist, however, the farmer, the miner, the railroadman and the factory worker had difficulty surviving the long-term deflation to which the country was subjected. Farmers had no way of forming a pool or a trust to get a decent price for their production. Wage earners, even with active union support, had little leverage to get their wages raised. Dissatisfaction was most acute in the South and the West. In the Middle Border States, the economic bubble of rising prices for wheat and corn, land speculation and railroad building of the early 80s had burst as a result of the drought of 1887. Between 1889 and 1893 over eleven thousand farm mortgages were foreclosed in Kansas alone. The situation in Nebraska and Iowa was almost as bad. On the industrial front more strikes occurred in 1890 than in any previous year in the 19th century.[3]

Out of this unrest came the Populist party, a political movement of revolt. It drew recruits from the Farmers' Alliances, Greenbackers, Knights of Labor and free silverites. The party was formally organized in 1891 and fielded a full ticket of candidates in the elections of 1892. The Populist platform demanded, among other things, the free and unlimited coinage of silver (an attempt to reinflate the economy); a flexible currency system controlled by the government not the banks; a graduated income tax; postal savings banks; public ownership of railroads, telegraph and telephones; prohibition of alien land ownership; an eight-hour day for labor; a ban on the use of labor spies; direct election of Senators and the use of the Australian (secret) ballot in elections.[4]

The Montana Populist party was formed in January 1892 at a convention in Anaconda with 230 delegates attending. Most of the adherents to the new party in western Montana were disillusioned Democrats. As the Populists organized to present candidates for city, state and national office it was evident that the Democrats would be the big losers. This presented Marcus Daly with a dilemma which he handled badly.[5]

When the new Populists ran candidates for all the city offices in Anaconda, Daly created resentment and fear by openly pressuring the Populists candidates to renounce their new affiliation and run as Democrats. While he had a record of great tolerance for labor organizing and labor unions, Daly made it painfully clear that he would use his great power to break anyone in Anaconda who didn't follow his lead in the political sphere. This was both the beginning and the end of the Populist party in Anaconda politics.[6]

In Butte, where W.A. Clark controlled the Democrats on the city council, Daly threw his support to the Republicans in the elections of 1892, electing Lee Mantel, owner of the *Butte Inter Mountain* newspaper, as mayor of Butte. These maneuvers splintered the Democratic party in Butte and Anaconda and helped undermine the party's campaigns statewide. As a result, the Republicans won the Governorship and the state's lone seat in Congress. The state legislature was split 26 to 26 leaving three newly elected Populists holding the balance of power.[7] (In the 1894 election the Populists increased their number to seventeen.) While the state went Republican, Grover Cleveland, Democrat, won the presidency.

The free and unlimited coinage of silver was one issue on which both mine owners and ordinary mine and smelter workers were united. They wanted higher silver prices and believed that increased coinage of silver was the way to achieve it. Higher silver prices would mean higher wages for the worker and higher commodity prices for the farmer. When, in 1889 Washington, Montana and the Dakotas became states, followed the next year by Idaho, Wyoming and Utah, they all elected pro-silver candidates to the U.S. Senate. Thus western Republicans, in exchange for voting for a tariff measure they didn't like, were able to win passage of the Sherman Silver Purchase Act of 1890. This law obligated the government to purchase 4.5 million ounces of silver a month, practically the entire domestic production, at market prices. But the Sherman Act not only failed to raise the price of silver, it failed to increase the money in circulation or the price of farm commodities.[8]

It did, however, keep silver prices from dropping even lower. When the Cleveland administration repealed the Sherman Act in November, 1893, it brought about the closing of silver mines throughout the West. In Pueblo Colorado, the Guggenheims closed a huge new silver smelter that they'd opened just two years earlier. The twin blows of the recession and demonetization of silver brought thousands of new recruits to the Populists cause.

In 1894, gold miners in Cripple Creek, Colorado, went on strike for an eight-hour day. Their 130-day strike ended in victory because the newly elected Populist governor refused to use the state militia or national guard troops to break the strike. This was a victory not only for the local Cripple Creek union but for the newly-formed WFM as well.[9]

Serious labor conflicts were also taking place in other parts of the United States which Butte and Anaconda watched with apprehension. There was a major strike at the Carnegie Steel Company at Homestead, Pennsylvania in 1892 which culminated in a pitched battle between the striking workers and Pinkerton detectives hired by Henry C. Frick, president of the company. Later that year the *Standard* reported that twenty union freight handlers were forced by the Pennsylvania Railroad to choose between keeping their jobs or sticking to the union. Five chose the union and lost their jobs. The *Standard*, continuing its pro-union stance and probably voicing Daly's philosophy, editorialized on January 8, 1893, that "The nation has passed through a year sadly marred by strikes, railroad men, factory men at Homestead, miners in Tennessee and Coeur d'Alene.... The majority of union men are conservative citizens who have houses and families and who would not resort to a measure likely to result in a riot unless forced by tyrannical measures to do so. The fact that a man is a union man is not sufficient reason for discharge." [10]

But the *Standard's* view was not shared nationally. In 1894, the American Railway Union, under the inspired leadership of Eugene V. Debs, took up the cause of the striking workers of Pullman Palace Car Company. The result was a sympathy strike which paralyzed transportation throughout the North. By June 27, the railroad workers in Livingston, Montana went on strike and on June 30, the union's Butte local joined the strike. This caused some of the Butte mines to close due to a lack of fuel for their steam-driven equipment. The adverse effects of the strike were so widespread that it precipitated intervention by the federal government on the side of the Pullman company to break it. As a result, the labor movement suffered another disastrous defeat. In Montana the strike leaders were jailed and fined. In Chicago, Debs was condemned to prison, where he studied socialism and eventually became a nationally known figure.[11]

In the West, Ed Boyce, President of the WFM, undertook a new effort to bring all the unskilled workers west of the Mississippi into one organization, launching The Western Labor Union in Salt Lake City in 1898. Within a year Dan McDonald of the Butte Miners' Union was President of the new WLU which, under the influence of Eugene Debs, shortly became known as the American Labor Union. The Anaconda Mill and Smeltermans' Union was affiliated with the new federation from the time of its formation. On February 16, 1897, the Mill and Smeltermen proudly hosted a packed meeting in the Evans Opera House to listen to a speech by Debs, soon to become founder and leader of the new Socialist Democratic Party of America.[12]

The silver issue split the country along regional lines in the presidential elections of 1896. The silverites in the Democratic party gained control and cleverly co-opted both the Populist platform, including free silver, and the party. For president, they nominated William Jennings Bryan, with his stirring "you shall not crucify mankind upon a cross of gold" speech. Pro-silver Republicans bolted to Bryan. Gold Democrats formed a separate ticket, but threw their support to the Republican, McKinley. It now was not only the money issue, but who would control the government, the business interests of the East or the agrarian and mining interests of the South and West.

In Montana, "Free Silver" became the dominant cure-all proclaimed by not only the Populist party but of large segments of both the Democrat and Republican parties, as well. Marcus Daly invested heavily and provided leadership for a new national organization known as the American Bimetallic League and gave support to the Montana Free Coinage Association. Prominent figures in the Republican party such as Lee Mantel and Congressional candidate Charles Hartman of Bozeman enthusiastically joined the movement. Republican Senator Thomas Carter, who was also national chairman of the GOP at the time, was torn between loyalty to his party and popular sentiment at home, but finally supported McKinley and the party's gold plank, which earned him the sobriquet of "Corkscrew" Carter. At the Democratic convention in Chicago, the Montana delegation, headed by Sam Hauser, who knew Bryan well, gave unanimous support to Bryan and his stand on silver.[13]

At the local level, Populists joined with Democrats in presenting a single slate of candidates, immediately dubbed the "Popocrats". The Populist candidates for governor, lieutenant governor and secretary of state, Robert Smith, A.E. Spriggs and T.S. Hogan were endorsed by both parties with the nominations for lesser state offices going mostly to Democrats. The Republicans split down the middle and nominated two separate slates of gold and silver candidates. So certain of victory for state offices were the silverites in both parties, that they spent most of their time outside the state working for the national ticket. Hartman and Mantel campaigned in the Midwest and Robert Smith in California.

While the politicians campaigned, the mineowners poured both gold and silver into the coffers of the national Democratic party. It was alleged that Marcus Daly became the largest single contributor to the Bryan war chest. The *New York World,* of October 16, 1896, reported that the party had raised $300,000 in Montana, $100,000 from Daly, $60,000 from employees of the Anaconda Company and $50,000 from W.A. Clark. Daly's son-in-law, James Gerard, later wrote that Daly had contributed $300,000 to the campaign. Whatever the final amount of Daly's contribution, the funds pumped into Bryan's campaign from the West were no match for Mark Hannah's efforts for McKinley among the country's wealthiest families, which altogether raised $16 million.[14]

As had been expected, the "Popocrats" in Montana made a clean sweep in the November 1896 elections taking practically every major state office and piling up an overwhelming majority of votes for Bryan, 42,537 to 10,494. The bitter disappointment of Bryan's defeat nationally in no way affected his popularity among the Montana public, especially in Butte and Anaconda. When he visited Butte the following summer, he was received like a conquering hero, by one of the largest crowds ever assembled. Charles Eggleston of the *Anaconda Standard* wrote a satirical verse about it entitled "When Bryan Came to Butte" comparing it, among other things, to Julius Caesar's return to Rome and Queen Victoria's diamond jubilee. Bryan had breakfast at the W.A. Clark mansion before delivering a speech at the Butte racetrack to an overflow audience. He then traveled to Anaconda, which honored him by being the first city in the country to name a school for him. He had a meal with Marcus Daly and made another of his famous speeches to an enthralled Anaconda crowd.

The election of McKinley was a triumph for industrial and manufacturing interests. It assured that the weight of federal authority would continue to support big business and, whenever there was conflict, against organized labor. A strike in Leadville, Colorado that occurred while the election campaign was in progress followed the established pattern and ended in the usual way. The owners closed down the mines and later brought in non-union labor. When union members used strong-arm tactics against the scabs, the state militia was called in and the strike was broken. There would be no change under McKinley.

During these same years, business interests, accompanied by the missionary zeal of many Protestant churches in the United States, began to awaken the public's interest in foreign affairs. Growing U.S. industrial prowess, expanding exports and American investments in sugar in Hawaii and the Caribbean all increased the U.S. stake in developments beyond our borders. Thus the nation began flirting gingerly with thoughts, long dormant, of Manifest Destiny and empire. Captain A.T. Mahan's brilliant series on the history of sea power had prompted increased budgets for the navy. Henry Cabot Lodge was advocating for a strong naval station in the Caribbean and the need for a canal across Nicaragua or Panama.

All of these currents coalesced and were given focus by the Cuban revolution of 1895. McKinley had been elected on a platform supporting Cuban independence. Although once in office he tried to pursue a prudent and responsible course, public opinion inflamed by irresponsible journalism, emanating primarily from William Randolph Hearst's New York *Journal* and Joseph Pulitzer's New York *World*, pushed the nation over the brink on April 20, 1898 when the congress in a joint resolution authorized the use of armed force to liberate Cuba. This meant war with Spain not only in the Caribbean but in the Philippines, as well.

The President immediately called for volunteers and within days the ranks of the regulars was overwhelmed with 200,000 raw recruits. It was a popular war, an opportunity for the whole country to demonstrate unity and patriotism after the bitterness and division of the Civil War.

Anaconda, enthusiastic like every other small town in the country, raised two companies of volunteers before the end of April to be part of the First Montana Regiment, which would be sent to Manila. Anaconda Companies K and M drilled and practiced shooting in Sheep Gulch not far from the construction site for the new court house. Matt Kelly tells the amusing story that, after the volunteers had left Anaconda, a charity ball was given, the proceeds of which were to be sent to the Anaconda boys in Manila. Two Hundred and fifty dollars were sent to each company. While K company received its money, he says, "The soldiers of Company M are still awaiting their share."[15]

The war was over almost before it began. On May 1, one week after the declaration of war, Admiral Dewey steamed into Manila Bay with his Pacific squadron and without losing a man tore the Spanish fleet to shreds. The Fifth Army Corps landed in Cuba and quickly won three battles. The Spanish fleet in Cuba was caught leaving Santiago Bay on July 3, and in a few hours was completely destroyed. The celebrations that took place the next day all over the United States, made July 4, 1898 a day long remembered. The United States had demonstrated to the world that it was a power to be reckoned with.

But the defeat of Spain did not mean a quick return for American troops in the Philippines. Instead of granting the country its independence as we had pledged to do in Cuba, the U.S. occupied the Philippines and commenced putting down local insurrections aimed at gaining independence. The Montana regiment did not return for sixteen months. Casualties among the Anaconda recruits of Companies K and M were the highest in the regiment, 35% dead and 15% wounded. When the boys finally did arrive at San Francisco, they were notified that they were on their own, and would have to arrange for their own transportation home. The city of Butte took up a collection, raising $30,000, and offered to bring all the Montana volunteers home. Anaconda mayor, John Madden, encouraged his local committee to raise sufficient funds so that Anaconda would pay its fair share. The Montana Regiment won high praise for its service in the Philippines. In 1905, its flag was made Montana state flag.[16]

In Anaconda, noisy receptions, full of high spirits and plenty of drink, were held for the town's returning heroes. There wasn't a hall large enough to accommodate a single reception for them all, so one company was feted at Turner Hall and the other at the new Ancient Order of Hibernians building. Music was provided by the AOH Fife and Drum corps.

Later in the week, Civil War veterans of The Phil Sheridan Literary Club held a meeting with the returning volunteers to form a new veterans' organization, "The Veteran Volunteers of the Spanish American War". After the usual round of speechifying, 54 veterans were signed up that evening.[17]

15. A New Century

Even though the United States declared its political independence in 1776, it remained an economic and financial colony for British and other European investors until the opening years of the twentieth century. The Montana Territory reflected this dominance, first in the fur trade and later in the large British holdings in cattle and sheep in eastern Montana. While the discovery of gold and silver in the West created immense new wealth for the young republic, and individual fortunes were made not only in gold but in steel, oil and railroads, the United States remained a debtor until the end of World War I.

But it wasn't hard to see, once the Civil War ended, that this vast country of immeasurable resources, with a burgeoning immigrant population and a spirit of unfettered enterprise would soon rival continental Europe and even Britain in industrial production. Before the nineteenth century was out, Anaconda had become the world's largest producer of copper and Butte hill, with its yielding up of gold, silver, copper, zinc and lead had justified Daly's boast as the "richest hill on earth". Similar records were set in other parts of the country in petroleum, iron and coal production. By the 1880s more tonnage was passing through the locks of the Saulte St. Marie canal on the Great Lakes than through the Suez canal. By 1900, the United States was producing as much coal as Britain and Germany combined and its iron and steel production surpassed all of Europe.[1]

Growing production led to rapidly expanding exports, so that 1892 was the last year in the nineteenth century that the United States had an unfavorable balance of trade. Leading American journals carried articles commenting on the decline in British industrial leadership and speculating on the eventual transfer of the world financial center from London to New York. In 1886 Andrew Carnegie had said "The old nations of the earth creep along at a snail's pace, the Republic thunders past with the speed of an express."[2] It was in that spirit of great confidence and heady optimism that the United States lived out the final years of the nineteenth century and rang in the new—the American Century.

8. The Deer Lodge County Courhouse shortly after it was built in 1900.

Anaconda showed its own confidence in the coming century by a steady stream of investment in new residential and business construction. The Standard Fire Brick Company had its best year ever in 1895 supplying brick for buildings in the town and at the smelter. The many new buildings, or blocks as they were called, in the commercial center of town, bearing the names of their owners, Barich, Davidson, Durston, Dwyer, Beaudry, Fortier, Sheilds, Dewey, Evans, Petritz, Stagg, Parrot and Flood, left no doubt that many of the towns leading citizens felt great confidence in its future. In addition, as Helena began construction of the new state capitol, Anaconda was building a sober and stately new edifice of its own for official business, the county courthouse.

Deer Lodge County was one of the original nine counties established by the Montana territorial legislature at Bannack in 1865. It was 70 miles wide and 250 miles long stretching north from the Big Hole River to the Canadian border. After the placer town of Silver Bow was abandoned, Deer Lodge became the county seat. This is where the original city plat for Anaconda and other legal registration for its founding was carried out. As new settlement took place,

new counties were formed out of the original area. In 1881, Silver
Bow County was created out of south eastern Deer Lodge County,
with Butte as its county seat. By 1889, when Montana became a
state, there were sixteen counties. In 1894, Granite County, with
Philipsburg as its seat, was carved out too, thus transforming Deer
Lodge County into a long thin entity running north from Gregson Hot
Springs to present day Glacier County, with Missoula County and
Granite on the west and Lewis and Clark on the east.[3]

After Anaconda lost the competition to become the state
capital, pressures grew to make it the county seat. Since most of the
county's population lived in or near Anaconda, the people voted in
1896 to move the county seat there from Deer Lodge. Soon thereafter
work began on the new county court house, one of the last important
public buildings to define the city's center. It occupied a commanding
site, against the foot of Burnt Hill, straddling the southern end of
Main Street looking down past the new Hearst Free Library, the Margaret Theatre, the Central School, the *Anaconda Standard*, the Montana Hotel, the Daly Bank & Trust Company, and the Copper City
Commercial Company to the Butte, Anaconda and Pacific Railroad
Depot, which similarly straddled Main Street at the northern end.

The town was still growing, the population having reached
12,000 by 1896. A steady flow of immigrants from Ireland and a
rising tide of Southern Europeans arrived to fill the new jobs created
by continuing expansion in the mines and in the smelter. Both the
Upper and Lower Works steadily increased their copper output
operating twenty-four hours a day. Lots were sold and houses built
in the town's 180-acre new addition of sixty-six blocks east of Ash
Street, which later came to be known as Goosetown.

Also a number of new churches were going up. As the
number of immigrants from the Austrian provinces of Slovenia and
Croatia increased, pressures grew to build another Catholic church in
the eastern addition for them. In 1897, George Barich once again
became active and, with the support of the Reverend H.B. Allaeys,
pastor at St. Paul's church, petitioned Bishop Brondell to authorize
a church for the new addition. With the Bishop's approval a committee was formed by the Reverend John B. Pirnat, assistant pastor
at St. Paul's and chaplain at St. Ann's Hospital, to raise funds and
procure property for the new church. Father Pirnat was a native of
Slovenia, but had been ordained in Helena in 1895 and said his first

mass at St. Paul's church where he was named assistant pastor. He canvased the city tirelessly to secure the financial backing necessary to begin the project. Marcus Daly's Township Company set aside lots on the corner of Fourth and Alder Streets for a church and a rectory. The church building was completed and formally consecrated in 1898 as St. Peter's, with Father Pirnat as its first pastor.

While it was initially known as the Austrian Roman Catholic Church, Bishop Brondel soon realized that it would also have to accommodate the Irish and French Canadians moving into the new addition. To achieve that objective he brought the gregarious and multilingual Belgian priest, Father Aime R. Coopman, from Livingston to take over the parish and assigned Father Pirnat to Coopman's parish in Livingston. After a short time, however, in 1901, Father Coopman was transferred to St. Paul's, as pastor, where he remained for many years. Father Pirnat returned from Livingston and once again became pastor of St. Peter's. Even though he spoke English, he gave most of his sermons to the integrated parish in his native Slavic tongue for the next fifty years.[4]

During this same period, the Swedish Lutheran church was built on the corner of Cedar and Fifth Streets, the Mormon church at 213 Oak Street, the First Church of the Disciples of Christ at 505 Oak Street, a new Baptist church on the corner of Fifth and Locust Streets and the First Methodist Church on Oak and Third Streets was enlarged.

The shift of the county seat created problems for ranchers, farmers and residents of small towns north of Deer Lodge who now had to travel long distances to transact official business. So in 1901, a new county was created from the area around Deer Lodge and north, with Deer Lodge as its county seat. Instead of allowing the new county to appropriate the name Deer Lodge, which coincided with the name of its county seat, the State Supreme Court ruled that the new county must be given a new name. Anaconda would continue to be the seat of Deer Lodge County, creating the anomaly that the town of Deer Lodge became the seat of Powell County, named after Captain John W.Powell, one of the pioneers of the region, while Deer Lodge County was confined to Anaconda and a limited area around it. Thus, in less than 40 years, Deer Lodge County was transformed from one of the state's largest to one of the smallest. From its original area had been carved Silver Bow, Powell,

Granite and large parts of Glacier, and Lewis and Clark, and smaller parts of Pondera, Teton, Flathead, Jefferson and Madison.[5]

Although Daly's town of Anaconda had become one of Montana's most important communities and continued to grow and prosper as the new century approached, it could not throw off its other image: that of being a mere satellite to Butte, the town where the ore came from.

Butte and Anaconda were alike in many respects, both were rough and ready towns. Both depended on copper and copper-workers for their existence, and while engaged in different aspects of copper production, they shared similar interests. As a result, even though rivalries developed, there was always a strong fraternal feeling between the two communities. Both towns were predominately Democrat and heavily unionized, but since Butte was larger and its unions had many more members, Anaconda usually followed Butte's lead. And it wasn't only in political and labor matters. Almost anything that happened in Butte was important to Anaconda. Consequently, the two cities and their histories were inescapably linked.

But in sports and civic activities young Anaconda worked hard to maintain an identity separate and distinct from her older brother, Butte. And with that fierce local pride and tenacity gained Butte's respect and cooperation to the point where over the years a certain symbiosis developed between them.

In the early days, Anaconda's July 4 celebration and parade was a major event always heavily attended and participated in by Butte organizations and residents. It wasn't just the spirit and liveliness of the Anaconda celebration that attracted Butte people, There was also the fact that many would still be recovering from Butte's own big celebration on June 13, Miners' Union Day, (which was also well attended by Smeltermen) and were, therefore, more than happy to close down for July 4 and go to someone else's party.

In sports Anaconda produced enough good teams and accomplished athletes over the years to become indispensable to Butte-organized independent and semi-professional inter-city baseball, football, curling and hockey leagues adding sparkle and excitement to the encounters.

Butte, of course, was a much larger, more boisterous city, and unlike Anaconda at the turn of the century, it was not a company town. While Daly and his associates had assiduously bought up as

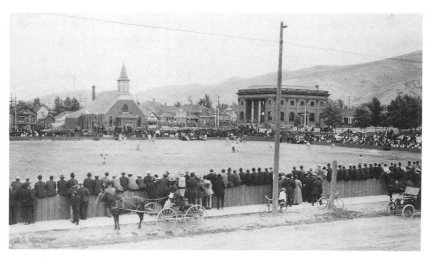

9. Baseball game on City Common. Note Hearst Free Library and First Presbyterian Church in Background.

many properties as they could around the Anaconda mine, other important producers were still in business on Butte hill, including W. A. Clark who owned thirteen mines there as well as the Butte Reduction Works, the Butte Street Railway, Butte Electric, Rocky Mountain Telegraph Co., Silver Bow Water Company and the *Miner* newspaper. In 1885, the Anaconda properties had produced 36 million pounds of copper. The other mines on the hill were close behind at 32 million pounds. There were also important mines, such as the Alice and the Lexington which were primarily silver producers. These were the competing interests in Butte as the new century began.[6]

Besides Clark, two large Boston banking combines were buying up claims in Butte: the Boston & Montana Copper and Silver Mining Company and the Butte and Boston Consolidated Mining Company. Other active companies were the Butte Copper and Zinc Company, the North Butte Mining Company, the East Butte Copper Mining Company, and the Butte and Superior Mining Company. Some of these companies operated small smelters in Butte. But in 1890, a hydroelectric plant and a smelter were built to process the ore from the Boston company mines in Great Falls, which had been linked to Butte by the Great Northern Railroad the year before.[7]

In June 1895, the Anaconda Mining Company was reorganized and became the Anaconda Copper Mining Company with 1,200,000 shares at a par value of $25 a share. Haggin was President of the new company and Daly was Superintendent. When Phoebe Hearst sold her shares in the new company to the banking house of Rothschild the following November, a new period in the Company's history began. Demand for Anaconda stock had been established and Daly was able to state in a letter to the *Anaconda Standard* that the sale price "makes the properties worth $30,000,000" and that subsequent sales of a portion of this stock had raised the Company's worth to $37,000,000. Later, the shares originally held by Tevis were also sold, so that just under 50% of the total shares were now in the public domain. The original partnership had all but disappeared. In 1899, Haggin retired as President and was succeeded by Daly.[8]

The immediate cause of Haggin's retirement was a proposal by the Standard Oil Company, which had been buying Butte copper properties and already owned a large block of Anaconda stock, to purchase majority control. Daly, whose primary interest had always been additional investment and expansion, saw the proposal as a way to access large sums of new capital for investment, so was in favor of going ahead. Haggin, who had always run his own show, wanted no part of this. Executives of Standard Oil, well versed in forming and controlling large industrial trusts, wanted to consolidate all the mining activity in Butte under one giant umbrella. This done, they could manage all aspects of the operation, control production, set freight rates through forced rebates and fix prices for finished copper. This was to be the first step in forming a national copper trust.

Henry H. Rogers, A. C. Burrage and William G. Rockefeller, principals in Standard Oil, had already bought up large blocs of stock in both the Butte & Boston Company and the Boston & Montana Company and were about to replace Daly's original backers, Haggin, Hearst and Tevis. With these changes, Butte and the young town of Anaconda were not only about to enter a new century but to start a new and very different chapter in their existence.

The Amalgamated Copper Company was organized April 27, 1899, and capitalized at $75,000,000. Haggin was paid $15,000,000 for his Anaconda holdings. Daly received $17,000,000 in Amalgamated stock. The original officers of Amalgamated were Rogers, President; Daly, Vice President; William G. Rockefeller, Secretary and Treasurer. As was usual in many such financial operations, the stock was generously watered. The nominal value of the new company was over twice the cash investment of $39 million of the principals. Thus copper joined oil, steel, sugar and tobacco as a primary commodity under the control of a giant trust and Amalgamated Copper became a familiar name on Wall Street, in the business community and in the halls of government.[9]

Once the deal was made, Amalgamated made typical "Standard Oil" maneuvers in the financial community not only to cover its investment, but to increase its capital without relinquishing control of the company. The initial investment of $39 million was covered by loans of an equal amount from New York's City National Bank. Then a public stock issue was floated accompanied by a publicity campaign designed to convince investors that they were getting in on the ground floor of one of the world's most sensational buys. The *New York Times* hailed the new stock offering as "the biggest financial deal of the age,". Great crowds gathered outside City National Bank to get in on this once-in-a-lifetime opportunity. It turned out to be the largest stock subscription in Wall Street history. It was so oversubscribed that Amalgamated issued only one share of stock for each five bid for. Thus the public paid $26 million, at $100 a share, for a one-third interest in the Anaconda Company assets, while Henry Rogers and friends held two-thirds interest in those assets for a net investment of only $13 million. During the months that followed, Amalgamated insiders began unloading some of their inflated stock and set off a selling wave that drove the price down to $75 a share, at which point they commenced buying again.[10]

16. *"What's the Price of a Vote Today?"*

But in Butte, Standard Oil type maneuvers would not render such quick results. The mining city was not about to give up its rambunctious, free-for-all ways without a fight, nor would it slip gently into the harness of a company town. It would only go kicking and screaming and creating as much confusion as possible. It was as though Daly, having signed a deal with the devil, had loosed the furies of Irish legend against everything he had spent his life building. The hero or villain of the piece, depending on one's point of view, was in many ways a personification of the city's many-sided and contradictory character. He was a handsome, debonair schemer, Frederick Augustus Heinz.

Actually, "Fritz" Heinz came on the scene even before Daly sold out to Rogers, Rockefeller and company. Heinz knew the copper business inside out, having started in Butte as a young surveyor for Boston and Montana. Later he worked on the editorial staff of *The Engineering and Mining Journal* in New York compiling copper statistics, and studied metallurgy at Columbia University's School of Mines. Having observed the gigantic fortunes amassed through predatory and callous manipulation of financial and commodity markets, he returned to Butte, still in his twenties, with an inheritance of $50,000 determined to challenge the copper kings in their own territory using any means necessary.[1]

Heinz concluded soon after his arrival that there was as much or more money to be made through the courts as through mining and processing copper ore. Just as Daly and Clark were manipulating the legislature to achieve their ends, he decided to try to manipulate the judiciary by exploiting the Apex Law of 1872.

The Apex Law held that the owner of a claim could follow any vein of ore that apexed (came to the surface) on his claim wherever it might lead. In Butte, with its faulted and complicated geological structure, it was difficult to establish where a vein apexed. Before a friendly judge and with testimony from well-paid expert witnesses, a convincing case could be made for practically any claim.

But Heinz was not only an unscrupulous manipulator, he was also a spell-binding stump speaker and a persuasive demagogue able to wield immense power at election time. He had little trouble convincing the miners, most of whom worked for Amalgamated, that Standard Oil was an octopus squeezing the life-blood out of Montana and the nation. He charged that "Standard Oil controls all the lamplight of the nation and now it is intent on controlling the electric light through manipulation of copper."[2]

He began causing trouble for The Boston Companies and the Anaconda Company in 1897, when he managed to elect William Clancy as one of the two district judges in Silver Bow County. Heinz had already bought and leased claims all over the district. Now using Clancy's friendly court, he began tying up Anaconda and the Boston companies' mining operations through surprise inspections, injunctions stopping all work, and levying heavy fines for not carrying out court orders. Clancy repeatedly rendered judgements that the veins of some of Butte's richest and biggest producing mines apexed on Heinz's property and therefore belonged to him. When these judgements were rejected or disputed, the obliging Judge Clancy issued injunctions to close down the mines in question.[3]

Heinz fight with Anaconda-cum-Amalgamated brought him into a natural alliance with W. A. Clark, who was still trying to get himself into the U.S. Senate and who welcomed the opportunity to cause his old rival, Marcus Daly, grief. They began a noisy and scurrilous campaign against Amalgamated in Clark's newspaper. Heinz bought another newspaper, the *Reveille*, to add to the noise.[4]

The Butte Miners Union, witnessing the open warfare among the mine owners, took full advantage of the situation to press its claims for improved working conditions in the mines. With Heinz on the stump making exaggerated promises to keep the miners happy, union leader, Dan McDonald, persuaded Clark that he would be the most popular man in Butte if he established an eight-hour day in his properties. Heinz immediately followed suit and the two of them noisily complimented the miners on their progressive spirit.[5]

At about the same time, at the convention of the Montana Democratic party in Anaconda, in 1898, Clark made his new bid for the Senate and set in motion a series of events that was to scandalize the nation. Clark Democrats won majority control of the party at the convention and in the November elections the Democrats won a decisive victory statewide. But Clark couldn't count on being elected by the legislature because Daly still controlled a bloc of Democratic votes. Confident that he had again stopped Clark, Daly left Montana for New York City on business right after the elections.

However, Clark had powerful supporters in his corner for this new challenge. Sam Hauser, John S.M. Niell, owner of the *Helena Independent,* newspaper and Walter Cooper, a powerful Bozeman lumberman had persuaded Clark that he must become Senator to stop Daly from taking complete control of the state. They were alarmed at the continuing expansion of the Anaconda company not only in Butte and Anaconda but beyond. Daly was in the process of expanding his holdings in the Bitter Root, buying out A.B. Hammonds lumber operations in the Flathead and around Missoula and making investments in coal mines in Belt and in building a new plant in Great Falls. On the political side, the Daly forces were strong among the Populists, who had rejoined the Democratic party in the elections of 1896, and with the silver Republicans like Lee Mantel. Hauser, who had allied himself with Daly in the past, now believed he had to be stopped. Clark, as Senator, could do it.[6]

Even before the legislature opened in January 1899, there were rumors that Clark was prepared to spend a million dollars to get elected and would pay $10,000 to any legislator who would vote for him. Charlie Clark, who was working hard to get his father elected was reported to have said, "We'll send the old man to the Senate or to the poor house." It's a sign of the times and of the place that it didn't occur to young Charlie that bribery was illegal and that a third alternative "for the old man" could have been "the penitentiary". Christopher Connolly, who chronicled the entire episode in his 1938 book *The Devil Learns to Vote* reports that businessmen in Helena argued that bribery was necessary and that the purchase of votes was discussed almost as freely as the weather. A morning salutation became, "What's the price of a vote today?"[7] But Clark could do no wrong in Helena, the city that he had made the capital.

The Clark forces first tried to buy off Daly Democrats. When Senator Fred Whiteside of Flathead County publicly produced $30,000 in crisp $1000 notes given to him by John Wellcome, Clark's attorney, to purchase his vote and that of three other Daly men, Clark's Butte *Miner* denounced it as "A damnable Conspiracy". The grand jury convened to investigate the charges was rumored to have shared in the same largess as the legislators and conveniently returned a verdict of insufficient evidence. The Clark money machine rolled on.[8]

Failing to pick up sufficient support from Daly Democrats, the Clark forces turned to the Republicans. Party stalwarts like old Wilbur Sanders were scandalized. Rumor had it that the price for a vote had reached $20,000 and there was at least one case where $50,000 was paid. This was at a time when a room in a good hotel cost $2.00 or $3.00 a day and the most sumptuous of meals not over $5.00; a good horse could be had for $75.00, a modest ranch for $2,000, and the annual salary of a U.S. Senator was $5,000.[9]

On January 28, when Clark finally won on the eighteenth ballot, only four of the nineteen Republicans had been able to resist Clark's golden handshake. The Daly forces claimed that Clark bought forty-seven votes for $431,000 not including the $30,000 that Whiteside had turned in. He also allegedly had offered $200,000 additional to legislators who refused him.[10]

Helena, once again as in 1894, went wild over Clark's victory and celebrated day and night at the old man's expense. The saloons poured free champagne for all takers confident that Clark would pick up the tab. Bonfires were lit in the streets and fireworks filled the frigid evening air. Clark, perhaps anticipating that he would have to continue his fight, even in victory gave no quarter and excoriated the wicked men arrayed against him, who had resorted to "treachery, falsehood, deceit, diabolical conspiracy..."[11]

The Daly forces lost no time in launching a counter attack. Whiteside, as a means to gather evidence for further use, brought libel suits against the *Miner* and the *Helena Independent* and initiated disbarment proceedings against Clark's attorney, John Wellcome, for bribing eight legislators. Clark's men immediately tried to suborn Judge William Hunt, offering him $100,000 if he would refuse to hear the case. After two other judges on the case and the Attorney General prosecuting the case all resisted Clark's monetary inducements, the trial proceeded and Wellcome was disbarred.[12]

In the meantime, Daly's men were gathering information to present to the U.S. Senate to prevent Clark from being seated. On December 4, 1899, the day Clark was to take his Senate seat, Montana's other Senator, and Daly's on-again off-again ally, Republican Thomas Carter, presented two petitions, one from the Governor of

Montana, "Popocrat" Robert Smith and other prominent Montana citizens, and another from anti-Clark members of the legislature. Both accused Clark of buying votes and asked the Senate to reject him. On December 7, the matter was sent to the Committee on Privileges and Elections chaired by New Hampshire Republican W.E. Chandler.[13]

For the first three months of the new century, the entire nation was treated to an endless parade of witnesses before the committee who described in embarrassing detail the depths to which political processes in Montana had sunk. Each new revelation made banner headlines in the New York and Washington newspapers. Clark, Daly and Hauser all admitted to having spent large amounts of money on the election, but claimed it was only for legitimate purposes. Clark denied knowing that his agents had offered bribes. He was heard to have said privately, however, that, "he'd never bought a man who wasn't for sale". In the end, the weight of the testimony, not only from those who were offered bribes and refused them but from many who had accepted them and couldn't explain how they'd come into such wealth, was overwhelming. The committee concluded unanimously that "William A. Clark was not duly and legally elected to a seat in the Senate of the United States by the legislature of the State of Montana". Faced with the certainty that the full Senate would follow the committee's recommendation to deny him a seat, Clark tearfully and angrily announced his resignation on the floor of the Senate May 15, 1900.[14]

Having thoroughly disgraced himself, his state and the many who accepted his largess, one might have expected Clark to quietly retire from the public eye. But this was not his way. He immediately began scheming anew. He lured the Governor of Montana, Robert Smith, to go to San Francisco as counsel in a trumped-up mining suit. While the governor was away, the lieutenant governor appointed William A. Clark to fill the vacancy created by William A. Clark's resignation. Smith, when he returned, called the appointment invalid and named Martin McGinnis to fill the vacancy. The U.S. Senate, faced with two appointments, adjourned without acting on either of them. So for the second time, Montana had only one Senator in Washington. It would take the election of November of 1900 to determine the legislature that would elect a new Senator.[15]

But by then the political scene had changed considerably. Heinz had now established himself in the politics of Silver Bow county and controlled the judiciary there. Standard Oil executives had become owners of the Anaconda properties and made an easy target for anti-Daly political charges. But most importantly, Daly was not in good health. In April 1899, he went to Carlsbad, Germany for treatment of what was reported variously to be diabetes, a heart condition and a kidney ailment. He was so sick at the Clark hearings in Washington that he was confined to his bed most of the time and the committee had to make special arrangements with short sessions and frequent breaks in order to get his testimony. He went to Carlsbad right after the hearings, returning in the fall even sicker than when he had left. When he arrived in New York he was taken to his private suite in the Hotel Netherlands, where he remained.[16]

Thus, unlike earlier such contests, when Daly was seen regularly on the streets of Anaconda and Butte, frequenting the local saloons, talking with his workmen, slipping a $20 gold piece to a miner down on his luck, inquiring about family members, this time Daly was not around. Instead, Heinz dominated the scene buying drinks for everyone and loudly criticizing Standard Oil and Daly's sell-out. He organized rallies at which one of the popular singers of the day, Cissy Loftus, sang to the tune of "Wearin' of the Green" an anti-Standard Oil ditty with the words, "We must down the kerosene, boys, we must down the kerosene." [17]

Of course, by this time Amalgamated Copper had a well-oiled political machine which went into high gear to again keep Clark out of the United States Senate. But Daly's presence was missed. The natural gulf between labor and management, which Daly had always easily bridged through his identification with his workmen, had grown immeasurably wider. An impersonal, out-of-state, giant corporation was taking over. Management decisions would be made on Wall Street in New York City, not on the spot in the mines or the smelter.

Heinz had a receptive audience among most residents of Butte and Anaconda. Amalgamated was no friend of the worker, it was the enemy; and this feeling was to become a permanent part of the popular culture. It was taken in like mothers' milk at birth and continually reenforced, generation after generation.

The Clark forces dominated the Democratic Party convention of 1900, and worked out a fusion ticket with the Populists and a platform which denounced Amalgamated, endorsed free silver and the eight-hour day, the exclusion of oriental labor and a vigorous national anti-trust policy. The Daly Democrats, clearly outgunned, formed the Independent Democratic Party which put up its own candidates including Amalgamated attorney and rising star in the Company, Cornelius Kelley, for Congress.

When the votes were counted that November, Heinz had elected his judges and Clark had a legislature that would put him into the Senate at last. The Clark Democrats honored their platform promises and introduced an eight-hour workday bill which was signed by Clark Democrat, Governor Joseph Toole, on February 2, 1901. Thus it came to pass that Clark rather than Daly presented the miners with that prize of prizes, the eight-hour day, making Montana one of the first states in the country to have an eight-hour law.[18]

However, this was for miners and construction workers only. The smeltermen in Anaconda still put in 13 hours on night shift and 11 hours on the day shift, seven days a week for $2.75 a day. It wasn't until 1906 that the Smeltermans' Union was able to negotiate an eight hour day, and break the twenty-four hours into three shifts instead of two, inserting an afternoon shift between the day and night shifts.[19]

17. *"The Mighty Oak Has Fallen"*

A week after the election, on November 12, 1900, Daly died in his suite in the Hotel Netherlands. After services in St. Patrick's Cathedral in New York City with Bishop Brondel of Helena officiating, Daly was interred in a mausoleum in New York City's Greenwood Cemetery. All Montana mourned the death of this remarkable man, who in twenty short years, had done so much for the state.

He was but one of many self-made men who became empire builders in the latter half of the nineteenth century. His name never reached the prominence of a Carnegie, a Morgan, or a Rockefeller. His fortune at the time of his death (reportedly $10 million) was small even compared to that of his rival W.A. Clark then said to be worth $50 million. But he had made an indelible mark on Montana. He was remembered not only for the copper industry that he had brought into being, but for his warm, friendly, jovial nature, for the innumerable people he had helped in times of crisis, and for his faith in Montana and its future.

It is ironic that this man who had made such a contribution to the state, who founded Anaconda, Hamilton and Belt, whose name was connected with everything of importance in western Montana, including numerous business ventures, a name which had loomed so large upon the scene during his lifetime, left nothing behind bearing his name. One of the tallest peaks in the Anaconda mountain range, along the continental divide southwest of the city, was named Mount Haggin by Daly for his partner, James Ben Ali Haggin, certainly a fitting tribute to a man who stood by Daly, like a mountain, against every adversity. His other partner, George Hearst, had the lake at the base of Mt. Haggin named after him as was Phoebe Hearst's island of culture in Anaconda, the Hearst Free Library. Even his Superintendent Mike Carroll had a village named after him. But Daly's name is strangely absent from the Anaconda he brought into being.

It would have been fitting at the time when Deer Lodge county was broken in two, that the part of which Anaconda was the county seat be named Daly County to honor him. There was an initiative in 1901 to do just that. State Senator T.F. Courtney from Butte introduced a bill to that effect into the legislature and it was passed unanimously. But the state supreme court ruled that the constitution intended to give stability to public processes when it prohibited special legislation "changing the names of persons or places". So Daly County never came into existence. Even more ironic is that for years the only reminder in Anaconda that Marcus Daly had once walked its streets was an elementary school that bore his name and the survival of a business arrangement, the Daly Bank and Trust Company, formed in 1901 by his wife Margaret, after the original partnership of Daly and Hoge was dissolved. A bank of all things, to remember a man whose legacy in money was a trifle alongside his monuments to industrial progress. Even the bank has long since disappeared.

But the posthumous absence of Daly's name tells us something very important about the man. In spite of his gregariousness, his hearty good nature, compassion and sensitivity, he was a very private person. People who knew him intimately admitted to knowing very little about what motivated him. He kept much to himself. To some he was a saint and to others, like W.A. Clark, he was the devil incarnate. Unlike Clark, who had gone to college and had taught school, Daly had little formal education, a fact which may have kept him from seeking public office as did Clark or George Hearst. He,like Hearst, was unpretentious to a fault, a characteristic which endeared him to his workers, but which limited his interest in public acclaim or in having his name attached to public monuments. It appears that his immediate family had a similar disinterest, since neither his wife nor his son and three daughters made any particular efforts after his death to see that his name be given prominence.

At the time that he died, Daly was engaged in the most extensive expansion he had ever undertaken of his smelting operations in Anaconda. He was building yet another new smelter, to be the largest of its kind in the world. Certainly this was an installation worthy of his name. It was completed in 1902, when his memory was still fresh in peoples minds. But he and his partners had decided on a name for it before he died. It would be the Washoe Reduction Works, recalling their glory days in Nevada where they had all gotten their start.

The Indian name Washoe, native to Nevada and having no relevance in the Northwest, had been inserted into the official Anaconda lexicon and it would persist. The city park which was about to undergo an expansion and a face-lift was named the Washoe park. Many years later, in the 1930s after the old Margaret theater had been replaced and that replacement burned down, a new state-of-the-art movie theatre was built in the latest art-deco architecture and design, to be, as every new building in Anaconda would always be, "the best in the state", Daly's name was once again ignored. The new movie house didn't even doff its hat to its predecessor the Margaret, but was named the Washoe. There was a short time in the 1960s, when this slight of the Daly name began to prey on consciences of the city fathers and Daly's most graceful architectural contribution to the town, the Montana Hotel, was given the name the Marcus Daly Hotel. That was before it was decapitated and gutted a few years later.

However, at the time of Daly's death there was an outpouring of grief in Anaconda, Butte, Helena and Missoula. Catholic services were held for him in all of these cities, and on the day of the funeral Butte was draped in black. Lee Mantel's Butte *Intermountain* expressed the magnitude of Montana's loss with the large headline: "THE MIGHTY OAK HAS FALLEN".

The New York Journal said: "There are countless other millionaires in the United States, but there is no man who can step into the shoes made vacant by his death and fill them." The Chicago *Record* commented: "...He was ever the friend of the working man...But withal, his eye had a glitter that stamped him as a leader of men...". The Minneapolis *Tribune*: "Daly will be remembered as the man to whom the West is most indebted." The Seattle *Post Intelligencer*: "There have been many great mining men reared in the United States but none of them has been quite what Marcus Daly has been."[1]

Butte, the headquarters of the Anaconda Company up to that time, would do more than Anaconda to preserve his memory. Company officials formed a memorial committee and commissioned America's outstanding sculptor, Augustus Saint Gaudens, to make a bronze statue of him. This statue was dedicated in 1907. The inscription on it reads:

"A pioneer miner who first developed the famous properties on the hill overlooking the site of the memorial which is erected by his fellow citizens in tribute to his noble traits of character, in grateful remembrance to his good deeds, and in commemoration of the splendid service he rendered as builder of the city of Butte and the State of Montana." [2]

The statue originally stood on Main Street not far from the gallows frame and entrance to the Anaconda mine. It was later moved to a site on the campus of the Montana School of Mines, now Montana Tech, overlooking the city of Butte.

With Daly's death, an epoch had come to an end. Daly's town of Anaconda would face the new century without the support and guiding hand of its founder and benefactor. But Butte had still not reconciled itself to becoming a company town and its votes for Heinz and Clark had thrown down a gauntlet which Amalgamated could not ignore.

Clark, who had finally achieved his long denied objective, was in no mood to do further battle with the giant taking over Butte hill. He became even more flexible when Standard Oil's Henry Rogers informed him that through Mark Hanna he could influence thirty votes in the U.S. Senate and again deny him his seat. Clark not only pledged his full cooperation, but agreed to sell additional mines in Butte to Rogers' Amalgamated Copper Company. In 1910, he sold his major copper properties in Butte to Amalgamated.[3]

It wasn't long before the hard-drinking, fast-talking thirty-one year old Fritz Heinz claimed for all those within earshot that, with Clark's defection, he was all that stood between the free-wheeling Butte of old and a buttoned-down, submissive one-company enclave. He would save the town from the octopus. He was already slowing production on many of Amalgamated properties through legal maneuvers in Judge Clancy's court and in 1900 managed to get Edward Harney, another friendly gavel, elected to the only other judgeship in the county. He was David fighting Goliath, but he had a mighty powerful slingshot.

Heinz didn't limit his war to the courtroom. He had his miners raid neighboring Amalgamated mines through underground cuts and haul ore out through Heinz properties. Such tactics sometimes resulted in hand-to-hand combat underground and many miners began carrying arms. When Amalgamated succeeded in getting court orders to inspect the mines being raided, Heinz men used unslaked lime, boiling water, dynamite and cave-ins to prevent entry. Tremendous damage was done to shafts, tunnels, timbering and machinery. Two Amalgamated workers were killed by dynamite when they connected through a cross-cut into a Heinz tunnel. It was later determined in court that Heinz, using his Rarus property for entry, had looted $1 million worth of ore from Amalgamated's Michael Davitt mine alone. For that offense he was fined $20,000 which Amalgamated never collected because the company Heinz formed to remove ore from those mines had no assets.[4]

But Heinz was out for bigger game. He was determined to break Amalgamated's back. In 1901, he planted agents in two of Amalgamated's subsidiary companies as stockholders who then brought suit to put those companies into receivership as part of an illegal trust. After sitting on the case for two years, Judge Clancy ruled that Amalgamated was doing business in Montana in violation of state law. But this time Heinz had gone too far. Within hours of the decision Amalgamated struck back with a vengeance.[5]

On October 22, it closed down not only the Butte mines but the smelter at Anaconda, the refineries in Great Falls, the lumber operations in the Bitter Root and near Missoula, the coal mines at Belt and Diamondsville, in effect all of its operations in Montana, throwing 20,000 people out of work at a single stroke. Six years earlier Judge Clancy had barely escaped being strung up by an angry mob of miners protesting his closing of three of the Anaconda Company mines with one of Heinz's injunctions five days before Christmas. How could Heinz now face the wrath of 20,000 workers?

This time Heinz took the heat himself in a magnificent demonstration of skill and courage. In Butte before an angry crowd of gun-toting miners, he coolly pulled out all the rhetorical stops on an old speech against Standard Oil that he'd polished in the 1900 political campaign. He ended with: "The same Rockefeller, the same Rogers are seeking to control the executive, the judiciary and the legislature of Montana." Then after a pregnant pause he added, "I defy any man among you to point to a single instance where I did one of you a wrong. These people are my enemies, fierce, bitter, implacable. But they are your enemies, too. If they crush me today, they will crush you tomorrow. They will cut your wages and raise the tariff in the company stores on every bite you eat and every rag you wear." Another pause and, "They will force you to dwell in Standard Oil houses while you live, and they will bury you in Standard Oil coffins when you die."[6] Fritz knew his audience. The suspicion against the giant from the east was universal. They all knew in their hearts that every word he uttered was true. How could they blame this man who'd fought the giant at every turn. Sure he fought dirty, but so did they, sure he was a scoundrel, but so were they. He fought for them in a way the union couldn't. Hadn't he helped them get the eight-hour day? He was the only one who had stood up to the Standard Oil steamroller.

But Amalgamated held the cards. When the shutdown was announced the price of copper immediately advanced one cent on the London exchange. Amalgamated had earlier accumulated over one hundred and fifty million pounds of copper in an effort to keep the price at seventeen cents. It could not only withstand a long shutdown, but could make money on the deal as well. This was widely commented on by the international press. George Hearst's old newspaper, the *San Francisco Examiner* reported: "Seventy-five million dollars were made today by the Rockefeller group of capitalists in the manipulation of Amalgamated copper stock." The London *Times* said: "It is believed in New York that the Amalgamated Copper Company will benefit by the Montana shut-down, as it will be able to sell on a rising market the large surplus of copper it has accumulated." [7]

Fritz was a gambler, but this time he'd lost big. He could still sway an audience and gain their genuine sympathy, but he couldn't put 20,000 men back to work. Neither could the governor nor the legislature nor the judiciary. Only Amalgamated could do that. As the days passed it became clear to everyone in the state where the power lay and who made the ultimate decisions. It was a demonstration of raw power, so overwhelming and so blatant that no one in or out of the state could miss its message. Copper ruled the state and Amalgamated owned the copper. It was a lesson not to be forgotten.

Governor Joseph K. Toole, a Clark Democrat who had run on an anti-Amalgamated platform, was notified that Amalgamated wanted a new law enacted immediately which permitted a change of venue in cases where there was a suspicion that the trial judge was prejudiced or when there was reason to believe that a fair and impartial trial could not be held in his court. The governor, clearly under duress (since he had vetoed a similar bill earlier), called a special session of the legislature on November 11, 1903. The Fair Trial Law was enacted and the governor signed it. The next day all of Amalgamated enterprises were reopened.[8]

Heinz continued to be a thorn in the side of Amalgamated for another three years but his hammerlock on the judiciary had been broken. As if to underline this, in the 1904 elections his two judges, Clancy and Garney were defeated. In addition, Amalgamated routinely began shifting venues on all of its important cases. Just as regularly, Heinz used whatever outrageous ruse available to him to make Amalgamated lawyers sweat. He employed gunmen, detectives and spies. Amalgamated responded in kind. The climate was such that practically everyone in Butte connected with mining carried a pistol. But since Heinz had failed to break the back of his opponents, he was now playing a game for more modest stakes, he was looking for a buyout. Butte's days of feuding mine owners was coming to an end. Butte would become a company town after all.

In February 1906 Amalgamated formed a new company, The Butte Coalition Mining Company, incorporated specifically as a vehicle to buy out Heinz. He was paid $10.5 million for all of his holdings. Immediately, one hundred and ten suits pending in the Butte District Court involving claims of more than $70 million were dismissed. Heinz left Butte for New York City where he formed a company, The United Copper Company, to deal in minerals, started a stock exchange firm and bought a bank. Within a year, mass sales of United Copper stocks destroyed the company's standing, caused a run on his bank and the suspension of his brokerage firm. The run on the Heinz bank had an immediate effect on related banks, such as the Knickerbocker Trust, one of the largest financial institutions in America, which collapsed, bringing on the Panic of 1907, with plunging stock prices, plant closures and bank failures.[9]

It was alleged at the time that Standard Oil forces, through City National Bank and J.P. Morgan, had manipulated the sale of United Copper stock and driven Heinz out of business. Heinz and his partner were indicted for fraudulent banking practices. His partner was found guilty and served time in a federal prison. Heinz was acquitted and returned to Butte for a short time but with little fight left in him. He died at age 45 of cirrhosis of the liver in Saratoga, New York.

18. Washoe Smelter

By 1887, the production of copper ore from Butte hill had surpassed that of Lake Superior making Butte the largest single source of copper in the United States. By 1890, Montana was producing 50% of all the copper mined in the U.S. The Anaconda Company was a leader in supplying the world copper market and introducing new production technology. As world demand for copper increased, so did Anaconda's output. New mines were opened on Butte hill and new veins, just as rich as some of the original ones, were discovered in existing mines. Daly and Haggin clearly meant to keep expanding their operations to meet the ever growing demand.[1]

They continued to modernize and increase capacity at the two Anaconda smelters throughout the 1890s. In 1892 they pioneered a new electrolytic process for refining copper which improved its quality and increased demand still further. But the two existing smelting plants were not large enough to keep up with the increased volume of ore coming out of Butte hill. So in 1898, they decided to build a new and larger plant with the capacity to concentrate and treat at least 5,000 tons of Butte ore a day. The most the Upper and Lower Works combined had been able to handle under optimal conditions was 4,000 tons.[2]

It was Daly's aim to build the largest and most modern reduction works in the world and with it to take the lead in the production and processing of copper. He had always employed the most advanced methods and machinery in his operations, recruited the most talented and experienced professionals for his staff, and encouraged experimentation and innovation in all phases of production. Now he wanted to consolidate advanced techniques and innovations in a completely new reduction and smelting plant. His aim was not only to process more ore, but to get higher yields utilizing a higher volume of lower grade ores.

A site was chosen on the southern foothills east of the town, three miles across the valley from the old works. The two old facilities, already the largest such installations in the world, would be

10. The Washoe Smelter 1901 before the first big stack was constructed.
Note old works across the valley with smokestacks still intact.

closed down once the new plant began operation. This was the vision
and dream that impelled Daly to make his deal with the moguls of
Standard Oil. He needed new capital and lots of it, more than his
existing operations and old partners could provide.

As the new century dawned construction on the Washoe
Reduction Works began. To supervise the building of the new plant
Daly chose his superintendent in Anaconda, Frank Klepetko, a native
of Bohemia with a mining engineer's degree from Columbia
University. Klepetko had previously directed the building of the
Boston and Montana smelter at Great Falls.[3] Ground was broken on
September 20, 1900 and as work got under way there was a large
influx of labor into Anaconda. The engineering and design staff was
increased, over a thousand skilled workmen were hired and temporary
housing was erected near the construction site to accommodate new
arrivals. With the Upper and Lower Works operating at full capacity
and construction going ahead on the new works, Anaconda, for the
second time, became a boom town.

Construction on the new works, known locally as the smelter,
continued even after Daly died and the fight between Amalgamated

and Fritz Heinze heated up. On January 22, 1902, after fifteen months of work, the first train load of ore from the Mountain Con mine in Butte was dumped into the ore bins above the crushers of the new Washoe Reduction Works.

The new smelter was an enormous complex of buildings covering over one thousand acres, almost as large as the town itself. It was built on the top and side of the sloping foothills so as to take advantage of gravity in processing the ore, which was hauled by rail to the uppermost point to begin its downward journey and transformation from rock into metal.[4]

Visitors found the new facility a showplace of the most up-to-date technology in the world. After the first big stack was added in 1903, all lit up at night it looked like a giant ocean liner floating above the town—a marvel of industrial progress where line after line of fifty-ton cars dumped ore into bins that fed a series of large crushing mills which systematically reduced the huge copper-laden rocks to dust. From there conveyor belts carried the pulverized stone through the concentration plant. On its way it passed through various screens, filters and classifiers then was mixed with water and chemicals to form a thick sludge. The water was brought in from lakes west of town via a wooden flume that snaked its way the length of the city along the mountains on the south side and into the smelter. The concentrated ore then went to roasting furnaces where fluxing materials, limestone or silica, were added and its sulphur was fumed out. The resulting calcine, a hot powdery white copper dust that flowed like water, was then transported in specially designed railroad cars to massive reverberatory furnaces and heated to a molten state where impurities in the form of slag were agitated to the top and "rapped off". The enriched copper matte was then run into thirteen-ton ladles. These ladles were transported by overhead cranes to the converter building where rows of pot-shaped, oxygen charged, flame-belching furnaces received the molten matte and melted out remaining impurities to produce a 98% pure product called blister copper. The blister was then electrolytically refined and molded into large slabs, "anodes", each weighing 465 pounds, to be sent to fabricating plants and turned into copper wire or other products.[5]

A month after operations started, work at the Upper and Lower Works began to be phased out. Only the silver mill and the electrolytic refinery continued operations. The houses of Carroll were carted off and transplanted in Anaconda, most of them in the new eastern division. The street car line was rerouted down Third Street traversing the length of the town and winding its way up smelter hill.

Within a year, Butte ore production had increased to 12,755 tons a day, and the new Washoe Reduction Works was expanded to accommodate this increase. Thus the smelting operation in Anaconda, originally built to handle 500 tons of ore a day had expanded, in just over 20 years, to more than twenty times its original capacity requiring at least four trains a day hauling sixty five cars loaded with ore. And it would be further expanded again and again. In 1912 a leaching operation to recover additional copper from the tailings of the two earlier smelters was installed. In 1913 a collective flotation process was introduced into the concentrator and in 1914 a separate facility was built for manufacturing sulfuric acid from flue gases.[6]

This was the basic plant completed in 1902, which would be enlarged, altered and upgraded over the next seventy-eight years, but which maintained its original configuration and functions until 1964. Then a concentrator was constructed in Butte leaving the Anaconda concentrator to function intermittently usually during emergencies.

One other feature of the new smelter received little attention at the time but came to play a major role in keeping the Anaconda Company competitive and the town of Anaconda alive. This was its sophisticated research department which was the Company's primary facility for experimenting with new processes and testing new technology. This coincided with the founding of the Montana School of Mines in Butte in 1900, which would provide a continuing stream of new professional talent and would reenforce metallurgical and mining research.

Thanks to these new facilities Montana became a leader in mining and metallurgy innovation. Ore concentration as practiced at the Washoe plant was the marvel of the mining world for years to come. In the new smelter laboratories processes were developed to recover manganese from the Butte ores. Here the Company experimented with and produced aluminum from native clays; and it was here that treble superphosphate was invented and electrolytic zinc perfected. No longer was it necessary to send technicians abroad to learn the most up-to-date technologies or to go out of state to study mining. The new Anaconda smelter was state of the art and would attract professionals from around the world.[7]

As new the processes proved themselves in pilot plants in the research department, the Washoe Reduction Works was expanded to use them in its regular operations. For example, a separate crushing facility and concentrator for zinc ores was soon added as were electrolytic tank houses to complete the conversion of zinc ores from rock to metal. Later came a separate phosphate plant where phosphate rock was treated with sulfuric acid to manufacture a high-grade fertilizer called Anaconda Treble Superphosphate.

During WWII the Reduction Works expanded yet again to produce manganese nodules needed for the manufacture of high-grade steel. A new concentrator and kiln were built in 1940 to process manganese ore from the Emma and Travona mines in Butte. Because of the enormous amount of electric power required to convert alumina-bearing ores to aluminum, work was pursued, not in Anaconda, but near the ACM power source in Great Falls. Eventually an entirely separate facility was built in Columbia Falls, Montana, using power from a new federally constructed Hungry Horse dam.[8]

Such a large and diverse industrial operation required a host of specialized departments to service the plant on a continuing basis. There was the BA&P railway which transported the ore from Butte to Anaconda; a foundry which manufactured tons of diverse iron and steel articles used in the production process; a brickyard to meet the large and continuing demand for building materials; a large power-house with enormous steam generators built originally to supply electricity to the entire operation and continued later on to supplement power brought in on high-tension lines over long distances from Great Falls; and a street railway department to transport workers from town to the Reduction Works. And, of course, there was a general office which kept the accounts and personnel records, made the purchases and administered the payroll. In addition, there were specialized crafts such as the painters, carpenters, masons, electricians, blacksmiths, ironworkers, welders, pipefitters, boilermakers, tinworkers, riggers, leadburners, firemen and watchmen who were assigned as needed throughout the plant. Finally, there was a security force, a safety division and a large maintenance department called "the surface", which did repair work on all the ACM properties in the Anaconda area from Georgetown Lake in the west, to Washoe Park and the city common in town, to the slum ponds beyond the smelter in the east valley.

In all there were sixty-six different departments, most headed by a superintendent, the entire operation presided over by a general manager. The total number of people employed by the Company in this enterprise varied depending upon the level of production at any given time, but was usually in the neighborhood of three thousand. This included a supervisory staff of about 200 men made up of the general manager, plant superintendents and shift foremen; a technical staff of over 50 metallurgists, chemists, engineers and draftsmen, and over 100 in the general office. (See statistical table below)

ANACONDA REDUCTION WORKS
Average Work Force 1905 - 1925 Inclusive[9]

YEAR		Average Number Employed	YEAR		Average Number Employed
1905	Jan-June	1,911	**1916**	Jan-June	4,952
	July-Dec	1,875		July-Dec	4,219
1906	Jan-June	1,961	**1917**	Jan-June	3,944
	July-Dec	2,288		July-Dec	3,350
1907	Jan-June	2,378	**1918**	Jan-June	3,640
	July-Dec	1,863		July-Dec	3,481
1908	Jan-June	2,095	**1919**	Jan-June	2,702
	July-Dec	2,219		July-Dec	2,581
1909	Jan-June	2,169	**1920**	Jan-June	2,751
	July-Dec	2,123		July-Dec	2,434
1910	Jan-June	2,090	**1921**	Jan-June	2,060
	July-Dec	2,039		July-Dec	889
1911	Jan-June	1,837	**1922**	Jan-June	1,869
	July-Dec	1,920		July-Dec	2,193
1912	Jan-June	2,093	**1923**	Jan-June	2,437
	July-Dec	2,270		July-Dec	2,541
1913	Jan-June	2,461	**1924**	Jan-June	2,529
	July-Dec	2,546		July-Dec	2,652
1914	Jan-June	2,682	**1925**	Jan-June	2,711
	July-Dec	2,118		July-Dec	2,626
1915	Jan-June	3,110			
	July-Dec	4,392			

At the beginning of the century, when the plant was completed, Butte and Anaconda had a reputation for paying some of the highest wages in the nation and workers, both immigrant and American born, were attracted to the two towns. Anaconda had the additional advantage of being considered a quiet, pleasant and orderly town, a good place to raise a family. Most homes had electric lights, running water and indoor plumbing, all of which were luxuries in those times. This was in stark contrast to the log huts and tar-paper shacks with no amenities, being constructed on the prairies of eastern Montana during those years by a growing wave of new homesteaders. A job on the smelter meant a definite improvement in living standards for most of the city's new arrivals.

However, there were negative factors connected with work in the two towns, as well. Anaconda was under an almost continual haze of smoke from its belching smelter stacks, with a pall of arsenic-laden dust settling on the surrounding hills and valleys. When the wind was right, residents could taste the sulphur in the air. And the high wages were more than offset by the dangerous and dirty work in the mines and on the smelter. At this time the estimated working life of a miner in Butte was not much more than ten years. While work at the new Washoe Smelter wasn't quite so dangerous, it still carried high risks. There was a potential for injury and death on practically every job on the "hill". This fact is brought home by the diversity of job-related accidents which might occur in the course of a day's work, from explosions and electrocutions to scalding by molten metals, burning by acids, crushing by machinery or railroad cars and falls from high places. Any one of these might easily cause the loss of a limb if not a life.

One of many stories illustrating these ever-present dangers involves a new hire being taught his duties by a fellow worker, "And in here, kid, you only get one chance, when the alarm goes off, you'd better get the hell out as fast as you can, or else." "Or else they'll fire me?" "No, or else they'll bury you!"[9]

These were the immediate risks of smelter work. The longer term risks were suffering and premature death from silicosis or cancer of the lungs, caused by working year after year on the "hill" breathing in the mineral dust, toxic fumes and chemical compounds.

An outsider visiting the smelter for the first time stepped into a totally new, confusing and even frightening world. After leaving the administration buildings near the gate at the end of the street-car line, no matter where you started, you would run across a jumble of rail lines, pipelines, hoses, and buildings that seemed to be thrown up helter-skelter in no apparent order. The buildings were made of steel and sheet iron or wood, with no insulation at all to keep out the cold winter winds. Workers performed their duties amid the hub-bub of a deafening hum from the power house generators and the constant noise from whistles, buzzers, air-hoses and torches mixed with smoke, dust and foul industrial smells. In later years as the smelter expanded, this feeling of chaos became even more intense. It didn't take long for a visitor to realize that this was dirty and dangerous work in a fearful place.

From the men on the high line freezing in the winter wind as they dumped the ore cars, to those scorching before furnaces in the reverbatory or the converters at the other end of the process, there was constant danger of injury or death. It was dangerous just walking around as a visitor, trying to stay out of the way of moving trains, cranes, chains, ropes and ladders while avoiding holes, piles of dirt, and live electric cables.

On the high line, exposed to the elements, ore cars with two pointed bins which opened at the bottom rolled in a constant stream stopping just long enough to release ore into the crusher bins. A crew of four men swung large sledge hammers against the sides of these cars to shake the ore out. They also had ten-foot-long steel bars to poke into the ore to loosen it. If they couldn't free the ore from below, one or two men climbed the cars and worked the bars from above, while the others pounded the sides of the car. In winter when the ore froze in the cars large torches were applied to melt the ice. The cars came all day and all night long, feeding the crusher. The four-man teams swung sledge hammers and worked their bars eleven to thirteen hours a day until 1906, when the eight hour day went into effect. In 1908, a new ore-dumping device called the tipple was installed which picked up the ore car and turned it completely over, dumping its entire contents into the crusher holding bins. This reduced the size of the crews on the high line, but there was still a need for some men to beat and poke the ore out of any cars where large boulders might have gotten caught crosswise.

After the ore was dumped, it passed to the crushers. This was one of the dirtiest, dustiest and noisiest places on the smelter. All the men working there wore muzzles over their mouths and noses and goggles over their eyes. On the giant first-stage gyrator crushers men were assigned to stand on a platform above the crusher mouths and pull out stray pieces of wooden mine stopes mixed in the stream of ore as it rained down into the machine. One misstep or an unusually stubborn piece of wood could drag a man off the platform and sweep him, along with the ore, into the crusher itself, an accident which, while rare, was almost always fatal. The noise was so great above the crushers that it was difficult to hear even the loud horns and whistles installed to indicate danger or breakdown.[10]

In later years when the big stack was built with its Cottrell treaters designed to catch the arsenic-laden flue dust before it went up the stack, an even dirtier job came into existence. It was called "rapping the treaters", and consisted of men wearing rubber gloves, dressed in acid-resistant woolen coveralls, with heads, faces and neck covered in gauze bandaging and face muzzles standing like strange beings from another planet on platforms and reaching through special openings with bamboo poles to tap the arsenic dust off of the long chains hanging inside. Each chain was charged with 75,000 volts of electricity which turned them into magnets attracting arsenic and mineral elements in the smoke as it passed up the flue. The current was turned off before the chain was rapped to release its accumulated dust. This dust was returned to the reverberatory for further smelting before it was finally treated to extract the acid. After a fatal accident at the treaters in 1928, when a man was electrocuted as he rapped a charged chain, safety doors were installed on all the openings to prevent access to the chains unless the current was off. Even so, a similar accident occurred ten years later taking the life of another unfortunate worker.[11]

The men in the concentrator worked in long, dark, foul-smelling, dusty buildings tending the belts that transported the crushed ore and the shaking screens which separated it, shoveling ore which had fallen off back onto the belts, oiling and greasing machinery which ran the belts and screens, adjusting the flow of water, oils and chemicals run into the crushed ore and testing for the required combination. Many men wore knee-high rubber boots because of the water and mud under foot and their work clothes were soon impregnated with the chemical reagents used in the flotation process. While the work here was not so dangerous as in the crushers, men were still injured or killed by getting caught in the belts and pulled into the wheels driving them.

The men in the roasters, reverberatory and converters sweated around ovens or before furnaces belching flame and spewing out smoke and toxic fumes. Anyone who worked in these departments for any length of time could recount instances of being scalded by molten metal and had the scars to prove it. And they were the lucky ones, others didn't live to tell about it.

One of the most dangerous places on the smelter was the tram railway that carried the red-hot calcine from the roaster to the reverb. This was one of the departments where the few blacks who worked at the smelter were assigned. They had to be on constant guard against being burned by molten calcine when explosions occurred as a result of blasts of cold air getting into the roasting furnaces.

For a visitor, the most spectacular sight on the "hill" was to see the giant converter pots belching multicolored flames that changed from yellow to orange to red and then to blue like oversized roman candles as they were charged with oxygen to intensify the heat. Overhead, men could be seen operating large cranes dumping ladles of molten copper into the converters or tilting the converter to pour off masses of molten copper or slag. Others men on the shop floor dressed in heavy leather aprons were responsible for "punching" the converters with long iron rods through special holes near the bottom to keep open passages for compressed air to enter the converter pots. Still others were "chasing" ladles, helping the craneman to position his giant steel hook precisely to catch the handle on the ladle prior to pouring. Directing these operations were the "skimmers", who determined when to skim off slag and when the copper was ready to pour by the size and color of the flames and way small particles of slag, copper and matte reacted to air blasts. The conversion of a sixty-five ton charge of matte usually required about three hours to slag off and another hour and three-quarters to finish.[12]

In any of these operations, a faulty piece of equipment, a broken gear or chain or a mistake in handling them would almost surely result in an accident. Deafening noise generated by the processes themselves made it almost impossible to hear warning signals when things went wrong. Men on the shop floor would sometimes miss a warning alarm when the overhead cranes began to pour the molten metal and would be hit and burned by splashes of it. Some died, others survived with major disfigurement, yet many returned to their old jobs as soon as they were sufficiently recovered.

In one of the powerhouses the noise from the large generators was so intense that the Company finally had to install a sound-insulated booth to provide some relief to the attendants between their hourly checks of the machinery. It was the human toll of such industrial processes that inspired journalist Hamlin Garland, writing about a Pennsylvania steel town in 1894, to comment, "Upon such toil rests the splendor of American civilization."[13]

The Company instituted a far-reaching safety-first program in every department on the smelter in 1918 and pursued it vigorously from then on. But, even with the most meticulous care, accidents were inevitable and sometimes the men questioned the Company's motives even in this area. Many a story circulated about a particularly cantankerous head male nurse who was around for years and who seemed much more interested in filling out the injury sheet than treating an injury. Some even averred that his interest in saving the company money was much more intense than in saving a life. He resisted sending for an ambulance in emergencies, waiting instead for a passing truck to take an injured man to the hospital. If a man were injured toward the end of a shift, he would make him wait in his office until the new shift came on to take him to the hospital just to avoid paying overtime to the ambulance crew.[14]

But as oppressive as was this combination of danger, dirt and noise, there were relatively few jobs left on the smelter that required grinding labor hour after hour or excessive muscle power to accomplish. With the possible exception of the masons, who still had to manually repair and reline roaster, reverberatory and converter furnaces while they were still hot, the Company's massive investment in machines and advanced processes had eliminated much of the backbreaking heavy physical labor of earlier days. The machines and the process had taken that over. In its place, were periodic bouts of hard labor alternating with routine and long stretches of tedium. Every job was different, but each had most of these elements on a daily basis.

While not every man tended a machine, it was the machines and the milling and smelting process that determined when and how hard one had to work. Some jobs allowed a man to work at his own pace, laboring hard and fast to finish in record time, then loafing the rest of his shift, or adopting a more leisurely approach and taking the full eight hours to finish. In many jobs emergencies and breakdowns determined how hard or fast one had to work. Newcomers arriving from a farm or other work backgrounds that required continual labor eight or ten hours a day would reach the initial conclusion that smelter jobs were easy and they didn't have to do any work to earn their daily wage. They learned with time that they were being paid not only to work, but to put up with the dirt, noise and danger involved in tending machines which did the heavy labor and set the work pace. Those who put a premium on sunlight and fresh air would conclude rather soon that such work was no bargain and would quit. Others moved on after suffering accidents or illnesses attributed to the work environment.

For those who stayed, this combination of elements created a special culture. Even those who accepted the danger, dirt and noise without protest realized that they had not made a particularly good deal and harbored a latent resentment against the Company. It was a job that kept food on the table and a roof over their heads, but they hoped in their hearts that their children would not have to do it. Many believed that since the Company was exploiting them they had a right to get even any way they could. This justified doing as little work as possible and encouraging others to do the same. Those who did more than the minimum were called stooges or "Copper Collars" and ridiculed. Newcomers learned fast that the price of acceptance was conformity.

While most workers had strong moral beliefs, few thought it was wrong to steal from the Company. "If the Company has it, why buy it?" was a popular saying. Most of the theft was minor in nature, whatever could be hidden in a lunch bucket. But from time-to-time major items would disappear. Truck drivers and others who carried equipment and supplies outside the smelter gates were particularly well placed for such thievery. The Company for its part seemed to accept theft as one of the costs of doing business. Only on rare occasions was action brought against anyone for stealing.

There was also a high tolerance among the men and even the foremen for drunkenness on the job, in spite of the danger involved. Alcohol, sad to say, was part of the mining tradition. It helped make a stark and dangerous life bearable. It gave a man courage to drop thousands of feet down a mine shaft and face the daily dangers of cave-ins and explosions. It offered relief at the end of a shift when he emerged soaking wet and exhausted. The saloon was an indispensable fixture in every mining camp in the country. It occupied a similar role in smelter towns, as well. In Anaconda there were two or three saloons on every corner of Third street east of Main, where the street car ran, to accommodate both the men coming off shift as well as those going to work.

Of course, most men did not go to work drunk. The smelter couldn't have operated under such conditions. However, some men did and with some frequency. Sometimes, if they were too far gone, they would be sent home. But often their fellow workers or even a sympathetic foreman would cover for them, while they slept it off in a corner someplace. Alcohol and alcoholism were, like danger and dirt, a part of the scene. And like them, they became objects of mirth as well as sorrow. A common and much told story concerns one very persistent imbiber. When told by his boss to go home and sober up, he answered, "It's no use, boss, she'll just send me right back up here again." [15]

In spite of it all, over the years the new Washoe Smelter set record after record for copper, zinc, phosphate and manganese production. It also became the center of life not only for those in on the start-up of the new facility but for three succeeding generations and their families. Despite the dangers inherent in their work, many who were there when it began survived thirty, forty, even fifty years of active labor before they decided to call it quits.

19. Coming of Age

As the new century started young Miss Anaconda, now seventeen years old, felt very good about herself. She had enjoyed nine consecutive years of growth and full-employment and looked forward to a prosperous future. She had completely escaped the ravages of the Panic of 1893, which devastated Granite and Philipsburg, closed all the silver mines in Butte and severely hurt the Montana economy over the next three years. The smelter did close down for five months in 1891, during Daly's fight with the Montana Union Railroad over freight rates. But once that was settled, the smelter fires were relit and production continued uninterrupted through the '93 Panic and into the twentieth century.

The courthouse building was completed early in 1900 and county officials began moving in on April 1. The official census for that year shows that Anaconda had a population of 9,453 and Deer Lodge county, 17,393. However, based on conflicting estimates from other sources, there is reason to believe that the census figure for Anaconda was understated. The Deer Lodge Trades and Labor council claimed that it represented 21 unions and 2,500 dues-paying union members. Another estimate puts the town's population at 12,163 in 1896. Publishers of the city directory arrived at this figure by multiplying the number of names in the directory by 1.75 to take into account the women and children who were not listed. In 1898, the same directory estimated the population at 15,095 by using a multiplier of 2.25, on the 7,267 names in the directory. If the same 1.75 multiplier had been used, the population would have been 12,717. While the rationale for the new multiplier was not explained, the influx of additional workers in 1900 to build the new smelter could well have put the total population near the directory estimate.[1]

Even though Marcus Daly was dead and Amalgamated now ran the Company and the town from its offices in New York City, the good will built up during the Daly years was still alive, if somewhat diminished by the Heinze-Amalgamated fight. The Company, whether by accident or design, named a series of general managers who endeavored to follow in Daly's footsteps in terms of making Anaconda a better place to live. These individuals were always the most important men in town, overshadowing the mayor and other civic leaders. After Frank Klepetko left to build more copper smelters in South America, E.P. Mathewson succeeded him. It was during these years that the Company sponsored a number of civic improvement projects.[2]

   As the population continued to grow, the construction of new houses and commercial buildings kept pace. The racetrack had acquired a new double-deck grandstand, and the Company had donated to the city the entire block on Main Street between Third and Fourth to be used for baseball games and other community activities. Called the City Common, it was graded level and fenced. It had a bandstand on the Third Street side, and a wire mesh baseball back-stop on the corner of Fourth and Main. It became an instant success with band concerts and baseball all summer long. Each winter, the entire block was flooded and turned into a skating rink and the bandstand was temporarily weatherproofed to become a change house for skaters. From November through February, a large Christmas tree, provided by the company, was strung with hundreds of electric Christmas lights and adorned the center of the Common.

   At about this same time, the City Park, which had been neg-lected, got renewed attention. It was enlarged, and the dam across Warm Springs Creek, which formed a small artificial lake with an island in the center, was rebuilt and provided with new spillways. During the warmer months of the year the lake was a popular spot for swimming, boating and fishing. A large dance pavilion was also constructed and a number of athletic fields for baseball and football were provided. The area was renamed Washoe Park and a spur of the street car line was extend to the dance pavilion. The park was fenced in and a circular roadway for horses and buggies was built around its inside periphery with a connecting entrance road off west Park Street. As the park expanded, a zoo was added which housed bear, buffalo, moose, elk, deer, antelope, ducks, geese and swans, and a small botanical garden was started. Matt Kelly tells the amusing story of a city council discussion about the lake, when someone suggested it would be nice if it had a gondola. One of the aldermen chimed in, "Not just one, there ought to be a lot of them. Why don't we get two and let nature take its course".[3] Later a fish hatchery would be added. Built in 1909, it was the first in the State.

In the town itself, the new century witnessed a flurry of building with the Ida Block going up on the corner of Commercial and Main streets and the Weiss Building, the Walker Building, the Parker Block, the Peckham Block and Thibidou Row in various stages of construction on east Park Street. Also, work was proceeding on the Silver Block on Hickory and Commercial streets and the Modern Woodmen's building on east Commercial.

By this time the city had two business schools, the Sisson Business College and the Copper City Commercial College. The Lincoln School, which burned down in 1897, was replaced on the same site in 1899 by a new, well-equipped three story, eighteen-room building "unsurpassed by any school in the state". In addition, there was the Bryan School on the east side of town and the Prescott on the west. A one-story brick building on the corner of Fourth and Main streets called the Central School had been replaced by a new two-story building, which, in 1891 became a combination primary and secondary school. Miss Lottie Blair was the first student to graduate from high school there in 1892. The following year the school had four more graduates and a growing number of students in each succeeding class. So, in 1902 with the total number of students in public schools exceeding 2,000, it was decided to float a bond issue and construct a new high school building. The architectural plan of M.D. Kern of Butte was selected by the school board over three other proposals.

The new Anaconda High School, built on Main Street at Fifth was completed in March 1903. With its air of restrained elegance it joined the Court House, the Hearst Free Library and the Margaret theater adding luster and grace to the southern end of the avenue. Besides classrooms, the new high school boasted a large library and assembly room as well as laboratories, rooms for typing and shorthand, and offices for the principal, superintendent of schools and the board of trustees. The Central School was then converted into facilities for teaching domestic science and manual training.[4]

During these same years, the Catholics in town were busy raising funds and to build their own schools. In the summer of 1898, Ursuline Nuns came to Anaconda to start a Catholic school. They bought a building at 1001 East Third Street, which had classrooms for girls as well as living quarters for the nuns. Shortly thereafter they bought another building across the street at 1014 Third Street which became a boys school. They next embarked on a building program, initiating construction of a large academy on the corner of Adams and Fourth Streets. The building, which was to have classrooms and living quarters for boarding students, would be constructed of quarried stone. Bishop Brondel came from Helena to officiate at the laying of the corner stone for the new building on August 30, 1901. However, the building was never finished because of some confusion about the nuns' rights to own property. The partially completed building was eventually torn down, and single family homes were built on the land.[5]

In 1903, the nuns had moved from St. Peter's parish to St. Paul's. Father Coopman, now pastor of St. Paul's, purchased land and constructed a new school for the nuns, St. Angela's Academy. It wasn't until 1907, that Father Pirnat had raised enough money to complete both a parish house for priests and a handsome new grade and high school for St. Peter's parish on Alder and Fourth street across from the church. In September, 1908, when the school opened for classes there were 463 students in the grade school and 21 in the high school. Eleven Dominican Sister were assigned by the diocese to teach there.[6]

The city directory for 1899 listed eight hotels, thirteen boarding houses, thirty-one rooming houses, fifty-five saloons, sixteen restaurants and cafes, eighteen grocery stores, seven meat markets, five livery stables, seven blacksmith shops, nineteen barbers, four dairies, twelve clothing stores, eleven churches, ten laundries (three of them Chinese) ten cigar factories and the Montana Meat Company's slaughterhouse at Mill Creek. There were also eighteen building contractors, seventeen lawyers, fifteen doctors (which included osteopaths, homeopaths and allopaths), five dentists, a hospital which could accommodate 22 patients in private rooms, and 105 in general wards. Anaconda was a town full of life and apparently little thought of death, for it had but two undertakers.[7]

In addition to physical improvements, the town was also experiencing the first signs of the technological revolution which the new century was bringing. Electricity and the electric light bulb were, of course, already taken for granted and had become an accepted part of the town's everyday life. But there was great fascination with another of Thomas Edison's inventions still in a primitive stage: the moving picture. One of the earliest exposures the town had to this novelty was a Veriscope presentation of the Corbett, Fitzsimmons fourteen round heavy-weight boxing championship bout which consisted of 143,007 frames and lasted two hours and thirty minutes. It was shown in Anaconda in November 1897.

In 1902, Smith's drug store on Main Street installed an Edison Peep-O-Scope. Subsequently, John Ward rented space on Park Street and treated the town to its first feature motion picture. Among his presentations were "The Mexican Bull Fight", "The Diamond Train Robbery" and the "Spanish Dancer". Then in 1907, D.C. Scott, a pressman at the *Anaconda Standard*, opened the Alcazar at 107 Main St. which presented a regular bill of films each week. In subsequent years Scott also owned and operated a number of other less prominent movie houses including the Grand Bijou, the Lyric and the Imperial. Other early movie houses were the Electric Theater which operated briefly on the corner of Park and Chestnut and Andrew Mandoli's Reel Theater at Park and Oak Streets. Later Mandole built the Bluebird Theater, which had a long life on East Park St. and was eventually absorbed by the Washoe Amusement Company.[8]

It was only a few more years before that other Edison invention, the phonograph, became a familiar household item. Sporting various brand names such as Gramophone, Graphola and Victrola, Anacondans saved their pennies and bought the fascinating machines that transmitted not only the human voice, but music and song captured on cylinders and later on black discs called records. They could buy records of Ted Lewis and his band or the famous Irish tenors, Chauncy Alcott and John McCormick, whom they had seen in vaudeville appearances at the Margaret Theater. It was, indeed, a wondrous age.[9]

Nevertheless, while these new contrivances were exciting novelties, they were still much too primitive to replace live music, singing, dancing and acting provided by visiting vaudeville troops. In the home the piano and sheet music still reigned supreme, and for parades and concerts the Anaconda Municipal Band under Frank Provost made the music. The band offered free summer concerts at the City Common and winter performances at the Margaret Theater and in the Evans and Turner Halls. Besides directing, Provost played the baritone horn and trumpet. A native of Belgium, he had trained at the Brussels Conservatory of Music and played in bands on trans-atlantic ocean liners before coming to Anaconda. He established a dry cleaning business in town and served as director of the municipal band for twenty five years. He later became Mayor of Anaconda in the 1930s.[10]

Another big change and a beneficial effect of Amalgamated's purchase of the Anaconda Company was to end of the policy of forcing company workers to buy at the Copper City Commercial Company. W.A. Clark and Fritz Heinze had turned this into a political issue in the election of 1898 and in Clark's fight for the Senate in 1900. There was even a move in the state legislature to prohibit companies from forcing their workers to buy at company stores. Heinze used it to illustrate the absolute control Standard Oil would have over everyone who worked for Amalgamated Copper. "They will cut your wages and raise the tariff in the company stores on every bit you eat and every rag you wear...and they will bury you in Standard Oil coffins when you die."[11]

Clark used it in defending himself against Daly's charges of bribery, alleging that Daly was a tyrant who exercised complete control over those who worked for him, even to the point of forcing them to make all their purchases in his company store. Clark, of course, didn't mention that the Copper City Commercial Company was only partly owned by Daly and that other partners in the enterprise were Daniel Hennessy and his own brother, C. Ross Clark.

Evidently Amalgamated had decided even before the company store became a political issue that it was in the copper business, not retailing. One of its first acts after taking over from Anaconda was to notify Frank Klepetko, general manager of the smelter, Frank Jones, superintendent of the BA&P and John Hickey, head of the foundry, that company workers no longer needed to buy at the Copper City. While this meant that Copper City solicitors would no longer roam freely over company property seeking out new hires and issuing them grocery passbooks, it by no means meant the demise of the Copper City. The store was a very successful enterprise with a large clientele, many of whom liked the convenience of being able to charge everything until payday. The Copper City was by then a local institution and continued to be an important part of the town into the 1920s.[12]

Even as late as 1896, fifty percent of the male population of Anaconda was single, living in boarding houses. In this regard, it was like Butte and many other western mining and smelting towns: wide open, hard-drinking, with non-stop gambling and a sizable red-light district, known as the "line". Some of its fifty-five saloons never locked their doors. All of the hotels and some of the boarding houses had bars, as well. With the influx of labor connected with the building of the new smelter and with the Upper and Lower Works running around the clock, the area of Main Street and Front and Commercial streets near the train depot presented a rollicking, turbulent spectacle both day and night.

Originally, the red-light district was concentrated around a parlor house called the "Globe" and ran for a full block along La Vita Street, which faced the railroad tracks, and around the corner toward Commercial Street on Hickory. The Anaconda Standard made sure that the town's citizenry stayed current on the latest developments there. Its Nov 30, 1889 reported: "FIRE ON THE ROW, Belle Riley's Disreputable House Wiped Out Of Existence On Hickory Street Between Front And First Streets". It went on to say that the Smelter City Hose Company and the Alert Hose Company had the blaze under control within a few minutes after they arrived on the scene and before the flames could reach the jail to its north and another house of ill-repute to the south. Others occupying the house were Lottie Lockwood and Ida Bates. The fire was blamed on "the carelessness of a Chinaman".[13]

Another article on February 17, 1893 reported: "Minnie Estelle, who lives in the badlands tried to carve her lover to mince-meat. She pushed the blade of a knife into the manly bosom of an alleged man who masquerades under the name of Fred Nash. Charged with disturbing the peace and quiet on First Street, she was fined $10." Another on the same day informed the town's citizens: "Minnie Williams was in judge Rockwell's court yesterday in a beastly state of jagosity. She was charged with conducting a house of ill fame and fined $1.00."[14]

A year later, the *Standard* observed: "The entire north side of La Vita Street will soon be adorned with a high board fence, thus obstructing the view of residents on the south side of that abbreviated thoroughfare and relieving passengers on the BA&P trains a disgusting sight..."[15] Appropriately, the city jail, a small stone building with an iron door and a small square iron-barred window, was on the east side of Hickory, just across the street from the "line".

By 1900, the city jail was in the new City Hall, the new BA&P depot stood on the site of a demolished parlor house and the "line" had moved across the tracks. It had not only moved, it had also grown and now ran for four blocks along Mainville Avenue from Hickory Street to Elm Street. When the police petitioned the city council to have an arc light put up on the corner of Hickory and Mainville to help reduce street crime, one of the Aldermen said he voted against such action because he "thought the denizens of that district rather favored darkness in which to pursue their avocation." There were four parlor houses in the district as well as dozens of cribs which prominently announced the fancy names of their occupants on the doors. Each girl paid a "court fine" at the City Hall every month.

The largest parlor house, the Monogram, was run by Madame Florence Clark, who drove a late model rubber-tired buggy and dressed herself and her girls in the most up-to-date fashions. She made sure that they always gave the local merchants a generous share of their business. Madame Florence even owned a number of race horses, one of which, Silver Stocking, was a record holder that raced at Salt Lake City and other western tracks.

Matt Kelly writes that soon after a leading Anaconda merchant married one of Madame Florence's girls, a large painting of a dark haired voluptuous nude that had hung over the bar in the old M & M saloon on Main and Commercial streets was quietly removed.[16] Those who found the Anaconda "line" too small, too tame or lacking in variety could always visit "Venus Alley" in Butte, which had the reputation of rivaling San Francisco's Barbary Coast and New Orleans' Corduroy Road. In addition to the hundreds of cribs and multi-purpose saloons along Galena Street, there were the more refined parlor houses on Mercury Street catering to wealthier clients. Almost one thousand girls paid the city fathers a regular $10 "court fine" to stay in business. It was said that on any given Saturday night as many as 4,000 men could be found milling around the area.[17]

20. Smoke Farmers

From the time the first smelters began operating, the smoke belching from multiple chimneys on the hills just across Warm Springs Creek was a problem. The roasting of sulfide ores produced a toxic cloud that hung heavy and low over the city. The Upper Works alone had twenty-six smokestacks. When the wind blew from the north or east, the taste of sulfur was in everyone's mouth. When it blew from the south or the west, heavy deposits of arsenic and sulfur residue collected on the vegetation of the farms in the east valley. The hills immediately surrounding the installations were already barren of trees, grass and other growth.

As copper production increased and the smoke got worse, farmers in the east valley began to protest. By 1895 they had joined together and brought suit against the Anaconda Company for damage to crops and livestock. This was the beginning of a series of famous "Smoke Cases" which attracted attention nationwide. While the company never admitted that the smoke had caused any damage to crops or livestock, it realized that it had a serious problem. It was leery of the courts, having already suffered at the hands of Fritz Heinze, and wanted to deal with individual farmers by buying their land and livestock and offering them jobs with the company. By 1902, it had settled with 102 of the 107 members of the farmers' smoke association, paying out $340,000 in claims. One of those who settled was Jesse Miller, a Civil War veteran and one of the early settlers along lower Warm Springs Creek, who sold his farm to the company and accepted a job as the company's first superintendent of parks. It was under his guidance that the new Washoe Park was expanded and beautified.[1]

The company also began to build higher smokestacks and to investigate other ways to make the smoke less harmful to the environment. But its troubles in this regard were not over. By November 1902, new complaints began to come in. Carcasses of dead horses and other livestock were found with increasing frequency scattered around the valley. One rancher complained that he had lost over a thousand head of cattle, another that seventy-five percent of his hay crop had been poisoned. A state veterinarian confirmed that the death of the livestock was due to arsenic and sulphur poisoning.[2]

Though the company denied that its smoke was poisoning the air, it was busy constructing a large new smokestack on the top of the smelter's highest hill. Thus, one of the features of the new Washoe Smelter was a huge 300 foot stack which was finished in August, 1903. This was intended to eliminate smoke damage to the surrounding areas. All the smoke from various individual operations was funneled through a series of flues running from each of them to a central flu and out this single stack. The flues served not only as funnels but as collection chambers and filters for dust particles and offending gases.

To make sure that everyone was aware of the magnitude of the company's investment in smoke control, the company staged an enormous public relations party on July 25, just before construction of the stack was completed. Tables to accommodate several hundred people at a sitting were set up for a gigantic luncheon inside the smoke chamber where the individual flues came together with the main flu. It was sixty feet wide and forty feet high and lighted with hundreds of electric light bulbs. Farmers from all over Deer Lodge Valley received invitations and a special train left Garrison at 10 O'clock in the morning and stopped at all stations along the way picking up passengers for the party. Residents of Anaconda were also invited for this celebration of the completion of the "largest stack in the world". The manager of the Montana Hotel, Dr. John S. Marshall, was instructed to prepare lunch for one thousand hungry guests. Elevators running inside the stack took anyone interested to the top to see the view of the city to the west and the valley to the east.[3]

It was the company's hope that its new stack would stop, once and for all, rancher's complaints and suits related to smoke damage. In addition, it began its own farming and livestock operations in the valley to demonstrate that no adverse effects were caused by smelter operations. But in 1905, the Deer Lodge Valley Farmers Association claimed continuing damages and approximately one hundred of them banded together and brought suit through a test case "Fred J. Bliss vs. the Anaconda Copper Mining Company". In the earlier cases most of the farmers and ranchers who had brought action against the Company had crops and cattle in the east valley fairly close to the smelting operations. This time many of the complaints came from ranchers as far away as Deer Lodge. It was said that the higher stack only spread the poison over a wider area.

The company's attitude had hardened considerably over its earlier reaction to such suits. Also, the initial action by the farmers' association was of such an extreme nature that the Company had to respond vigorously. The farmers sought an injunction to close down the smelter. Had this been granted, all the Company's other operations would have had to close also. Having demonstrated its strength in 1903 against Fritz Heinze, it was in no mood to negotiate with what it derisively called the "smoke farmers". It put together a high-powered legal team, headed by Cornelius Kelley of Butte, who would later become Anaconda's president and chairman of the board, and it started a noisy newspaper campaign against the farmers' suit. Daly's, *Anaconda Standard,* which had fought the good battle to make Anaconda the state capital, now was Amalgamated's paper and it turned its guns on the "smoke farmers".

It alleged that it was the Farmers Association's intention to close down the smelter entirely and thereby threaten the livelihood not only of everyone working there, but of the farmers themselves who depended upon Anaconda as a market for their produce. Much of the testimony came from farmers in the valley who had earlier settled with the company. They testified that while they had had crop and livestock losses before, they had not suffered any since the new stack had been built and they confirmed that if the smelter and the mines shut down they would not have a market. While the smelter-men knew that even with the higher stack there were days when the smoke over the town was almost unbearable, they were much more concerned about their jobs. Amalgamated had already closed down the smelter once to win its fight with Heinze. They wanted steady employment. Hence, popular opinion was with the company, even though there was a sense that the farmers were probably justified in their complaints.

The trial, which started in January 1906 and continued through 1907, was watched with interest and concern by all the smokestack industries in the West. Finally on April 26, 1909, after 27,000 pages of testimony had been recorded, Judge Hunt, accepting testimony from experts that preventive measures established at the smelter were effective, dismissed the suit.[4]

The *Anaconda Standard* reported: "Exultant blasts of the whistles on the Hill heralded to Anacondans yesterday's news of the decision. Brass bands, trailed by hundreds of boys, marched the streets and roman candles and rockets made the air brilliant."[5]

Even though the Company had won the legal battle, stories circulated for years after about the continuing effects of smelter smoke on plant life and livestock in the surrounding area. Horses were particularly vulnerable to arsenic poisoning from eating contaminated vegetation. Their mouths and noses would develop sores which would swell and inhibit their breathing. As breathing became more difficult or impossible the horses would fight for air, twisting around and around in a mad fit to the death, or breaking out of stalls and barns and running until they dropped dead. This was such a common phenomenon that almost every rancher living in the valley had his own story to tell about it.

Many years later, John Holtz, who was raised by his widowed mother at the turn of the century on a small dairy farm on Lost Creek just over the hill from the Old Works, recorded his own experience with what he called flu dust. He said that the cows could survive much better than the horses because they had long tongues and were forever licking their faces which washed the burning arsenic away. But "who knows what it did to their insides. Now, I know you're not going to believe this, but there are still a few of us around who can testify to it. When those cows died, if you cut off the lower jaw and boiled it, you could see that the teeth had become copper plated." He went on to say that there were still areas on Lost Creek where there were piles of rock left by the Company after it had run the contaminated top soil through screens before taking it to the New Works for processing to recover minerals that had been lost as flu dust at the Old Works.[6]

There was another story frequently repeated which illustrated the unusual lengths to which the Company went during the trial to convince everyone that crops were not damaged by smelter smoke. Large shipments of horse manure from Daly's horse ranch in Hamilton were shipped by rail and delivered in large quantities to certain friendly farmers, who were able to demonstrate within a short time unusually large yields of alfalfa.[7]

Also, while the trial was on, Deer Lodge County published an unusual booklet which praised the agricultural assets of the valley with photos of prize sheep, cows and beef cattle and extolled the yields from a number of farms which raised a wide variety of crops including hay, wheat, potatoes, beats, rutabagas, carrots and cabbages. It also mentioned the ACM experimental farm with 100 head of Hereford cattle and 640 acres of hay, grasses and root crops. That farm gradually expanded and came to specialize in sheep, regularly winning prizes at the largest agricultural fairs in the United States for outstanding pure bred rams and yews.[8]

While smoke was undoubtedly the most serious problem, there was additional poisoning of farm land in the valley from run-off mine tailings and leaching in the Butte mines which found their way into Silver Bow Creek and the Deer Lodge Valley. Early on, unwary farmers used this water for irrigation until they realized that rather than helping crops to grow it killed them. By that time the land thus irrigated had become useless for either crops or grazing. As mine production increased, the pollution of Silver Bow Creek became worse until the waters turned a sickly yellow in color and it became obvious to everyone that it was harmful to crops. By then those farming near it had changed its name. That clear stream which Conrad Kohrs had followed into the valley remarking on the abundance of fish and wild life and the tall grasses growing along its banks, now poisoned flora and fauna alike. That stream that once evoked thoughts of reflected sunlight as it arced in silver bows was now called Yellow Creek. Some called it Stink Creek.

All of this was taking place at a time when the President of the United States had aroused public opinion and was taking a personal interest in the conservation of natural resources. Theodore Roosevelt had dedicated his administration to the conservation of the nation's soil, streams, rivers, lakes and forests, and had set in motion a series of actions to put the brakes on the greedy and wasteful destruction of natural resources. Already, four-fifths of the nation's timber was in private hands, ten percent of it owned by the Northern Pacific, the Southern Pacific and the Weyerhaeuser Timber Company. Roosevelt began by setting aside 150 million acres of unsold government timber land as national forest reserve and withdrew from public entry an additional 85 million acres in Alaska and the Northwest.[9]

When he was informed that National Forest timber was being damaged by Anaconda smoke, Roosevelt, having fished and hunted near Anaconda, speedily filed suit against Amalgamated. This set in motion a prolonged negotiation between the U.S. government and Amalgamated. In 1908, the Deer Lodge National Forest was created with headquarters in Anaconda. It was carved out of parts of the Helena, Big Hole and Hell Gate Reserves. The Forest Service reported in 1909 that large stands of timber in the Forest were being killed by smelter smoke and again in 1916 that practically all the timber coming out of Mill Creek was dead. Arrangements were made for the Company to slowly and discreetly exchange timber lands that it owned in other parts of the state for those which the government claimed had been damaged. The last exchange was not completed until 1937.[10]

But some U.S. government actions against the Anaconda Company even preceded Roosevelt's conservation efforts. According to the *Northwest Tribune,* of Stevensville, Mont, Oct 19, 1900, in 1899, "John W. Griggs, attorney general in the McKinley administration, ordered U.S. Attorney W.B. Rodgers of Helena, to bring suit against the Bitter Root Development Company and the Anaconda Copper Company for stealing timber from public lands in Montana... The trespass involved the cutting of 23,525,128 feet of saw logs on unsurveyed lands lying within the limits of the Bitter Root Forest Reserve." U.S. Senator Thomas Carter of Montana pressured Attorney Rodgers not to file the suit. Rodgers, who had been appointed by Carter, complied. The *Tribune* alleged that "H.H. Rogers, the new head of the Anaconda Company...summoned Senator Carter to New York [with] Mark Hannah" and told them, "that unless the suit against Anaconda Company was abandoned, Mark Hannah would get no more money from the Standard Oil trust, the Amalgamated Copper trust and their affiliated companies. He told [Hannah] that in Montana he was spending millions to elect judges and a state ticket, and could just as well carry the state for McKinley, but that if Anaconda company were to be prosecuted for a million stolen government logs, McKinley and Mark Hannah should have no more money...Mark Hannah went to Washington and laid all the facts before McKinley...Thereupon the President undertook to straighten out the matter; it was fixed satisfactorily to Hannah, Rogers and Carter".[11]

But this wasn't the end of the matter. In 1906, Roosevelt reinstituted the suits. While Amalgamated was fighting the smoke cases, a grand jury returned 102 indictments charging that the Bitter Root Development Company wilfully denuded federal lands of $2 million-worth of standing timber. Rather than fight the government in court, Amalgamated decided to negotiate and was able to settle the suit by paying $156,000. In light of the alleged damages, such a small settlement was not only another obvious triumph for the new owners of Montana's copper wealth, but a clear sign that they had the wherewithal to easily pacify even a conservation-minded administration.[12]

In addition to these problems of an environmental nature, Amalgamated management was also aware that the mining and production of copper created serious long-term health problems for its work force, a dread disease commonly referred to as miners' consumption, known scientifically as silicosis. Both miners and smeltermen, who survived the risks of possible accident or death in the dangerous work they performed daily, faced the prospect of this debilitating and often fatal lung disease caused by long-term exposure to high levels of dust in the mines, mills and smelters.

Even though the Company was continually upgrading its properties, improving ventilating systems and introducing dust and smoke reduction measures, a large proportion of the long-term work force suffered in some degree from breathing and lung problems. It was so common, in fact, that it became accepted as inevitable. If you worked in the mines or on the smelter, you knew that eventually you would wake up in the mornings coughing violently and spitting up thick gobs of mucus, even blood as it got worse. It wasn't anything you went to the doctor about. He couldn't do anything and besides everyone you worked with had the same thing. It was part of the job. If it got too bad and you had trouble breathing, you'd have to quit work. While there were a few sanitariums that treated this condition, they were expensive and none was located in Montana. It was not unusual to hear someone mention a friend or relative who had "died of the con", referring to consumption. While there may have been a few who blamed the company for their condition, it never became a serious negotiating issue. Unlike the "smoke farmers" who organized to seek a remedy for their problem, the workers did not exert similar pressures on the Company with regard to lung disease.

Nevertheless, there were others in the state who were concerned. In 1906, the *Standard* ran a series of articles by Dr. Thomas Tuttle, Secretary of the State Board of Health, on tuberculosis, the high cost of treatment and the hardships suffered by people not able to afford the treatment. He suggested that this was a problem the state should address, that other states such as California, were providing treatment in state institutions and that Montana should do likewise. The *Standard* followed the articles with an editorial calling attention to the problem and suggesting that men of fortune might endow such an institution. Since Amalgamated owned the newspaper, the *Standard's* interest in the subject is puzzling especially its suggestion that such an institution be privately endowed. It is reasonable to speculate that the Company, possibly anticipating future lawsuits or pressure to fund a hospital for the treatment of silicosis, might decide to encourage the establishment of a state-run and-*funded* institution dedicated to the treatment of lung diseases. Whatever the motive, action was slow in coming.[13]

The state legislature, encouraged by Dr. Tuttle, took up the matter in 1909, but was able to pass only a measure providing for the teaching in public schools of the modes of transmission and methods of control and prevention for all communicable diseases. In 1911, Jim McNally, a native of Butte, roused by his own brother's death from consumption, worked with Dr. Tuttle and went before the legislature to present a bill that would establish a state-run and-funded sanitarium. The proposal called for $200,000 for buildings and $100,000 for maintenance. The bill passed but with the funding whittled down to $20,000 and $10,000.

In 1912, the state took over the operation of the Insane Asylum at Warm Springs, previously run by Doctors Mitchell and Mussigbord, purchasing from them the installations and much of their land. It was on this land, six miles northwest of Warm Springs, that it was decided to begin the construction of the tuberculosis sanitarium. The Anaconda Company, through its Vice President, Cornelius F. Kelley, donated $25,000 to the sanitarium fund and construction commenced under Dr. Tuttle's supervision. The new facility, consisting of a main building , six cottages for staff and a stable and chicken coop, was called Galen, not after the famous physician of Roman antiquity as many believed, but after Albert J. Galen, who was Attorney General of Montana at the time. Upon its completion,

Dr. Tuttle was named the first Superintendent of new state sanitarium. Thus, Deer Lodge Valley gained a new state-run institution, and all those suffering from tuberculosis or silicosis gained a sanitarium where they could be treated free of charge. The Company referred the worst cases from Butte and Anaconda to the new facility. Between 1913 and 1917 344 patients were admitted of whom 150 died. Most were between the ages of 31 and 40. As time went on and medical science advanced, the recovery rate for patients entering Galen gradually improved.[14]

21. Unions and Socialism

While Amalgamated's successes against Heinze and others gave it complete economic control over Butte and Anaconda, it still had not established absolute political domination and it could not shut out the currents of political ferment wafting in from other parts of the country. Since the 1850s working men in the United States had been hearing about Karl Marx and his recently published theories of class struggle. These views had gained a substantial following among German immigrants who, in turn, carried their countryman's message into the emerging labor movement. It had an instant appeal among immigrant laborers suffering the harsh working conditions in expanding industrial enterprises where there was little protection under the law against even the most blatant exploitation. It was through organizing these workers that the Marxists intended to advance their political agenda.[1]

Additional credibility was given to Marx's message by the rapid concentration of control over the nation's wealth and its means of production in the hands of a small group of bankers and industrialists around the turn of the century. Not only did individuals or families control entire states in the west, but by 1900, all the major railroads in the country were in the hands of six groups. The Morgan Belmont group controlled 24,000 miles of track, the Harriman group 20,000, the Vanderbilt group 19,000, the Pennsylvania group 18,000, the Gould group 16,000, and the Hill group 10,000 miles. Not satisfied with his share, Harriman was maneuvering to further reduce the number of players through a scheme that would have given him control of every important railroad in the country. J. P. Morgan, for his part, controlled banks, railroads, shipping companies and the production of steel, electricity, agricultural machinery, and rubber. By 1913, the Morgan, Rockefeller, National City Bank group controlled 341 directors on 112 corporations.[2]

The Sherman Anti-Trust Act of 1890, which was aimed at slowing down this concentration of power and preserving competition in the market place had proved completely ineffectual. It had neither stopped the growth of trusts and monopolies nor stamped out the abuses which accompanied it. There was no will within the executive branch to invoke the law and no disposition in the judiciary to support it. The Harrison administration filed only seven anti-trust suits, the Cleveland administration, eight, the McKinley administration, three, and the Supreme Court ruled unfavorably on those suits which reached it. As if in open defiance of the law, there were more trusts formed under McKinley than at any time in U.S. history. Instead, the law was turned against labor and was widely used to discourage the growth of industry-wide unions.[3]

It was in this climate that the socialists gained popularity and power in the labor movement. They soon dominated the Brewery and Bakery Workers, the Shingle Weavers, the Ladies Garment Workers, Fur Workers, the International Association of Machinists and the Western Federation of Miners. Also, at least half the members of the United Mine Workers of America were socialists, as were up to one third of the membership of the American Federation of Labor.[4] This was also the situation in Butte and Anaconda where substantial numbers of union members considered themselves Socialists.

In 1901 the Fabian Socialists broke away from the Socialist Labor Party and formed the Socialist Party of America, with Eugene Debs at its head. He had come out of the labor movement and was a strong advocate of the large industry-wide union to increase labor's bargaining power with management, rather than the craft union approach favored by the AFL. Debs became the Socialist's perennial candidate for President. While he never came close to winning, the increasing number of votes for him from 1900 to the beginning of World War I in 1914 gives some indication of the growing dissatisfaction with the two major parties in various sections of the country.

In Anaconda and Butte, which traditionally turned in large majorities for the Democrats, this disaffection, which had first manifested itself in the rapid rise of the Populists in the 1890's, was further exacerbated by the passing of ownership of the Anaconda Company into the hands of a distant and monopolistic Amalgamated Copper and the ongoing bitter battle with Fritz Heinze for control of Butte.

These currents produced a dramatic development in Anaconda in 1902. The Central Labor Council founded a new political party called The Deer Lodge County Labor Party which won a stunning victory in the fall, electing five of its six candidates to the State Legislature as well as capturing the county offices of sheriff, county attorney, treasurer and assessor. Shortly after this victory the Labor Party changed its name and became a local branch of the Socialist party of Montana with very close ties to the Anaconda Central Labor Council. On March 7, 1903 a confident and boisterous Socialist convention nominated John H. Frinke for mayor, Michael Tobin for city treasurer, Con McHugh for police magistrate and four candidates for the city council. The party's executive council later completed the ticket naming two more city council candidates. On April 6, with a turnout of 92% of the registered voters, Frinke, Tobin and McHugh were elected with comfortable pluralities along with three of their Socialist comrades to the city council. This was a startling and remarkable achievement for the Socialist party. It was its first victory in municipal elections west of the Mississippi. William Mailly, national secretary of the Socialist Party of America, telegraphed Frinke congratulating him for a "splendid victory for socialism."

Who were these Socialists and how had they swamped the traditional political parties? John Frinke was born of German parents in New York City. He was a cigar maker who owned his own shop and like another contemporary Samuel Gompers, founder of the AFL, was a long-time member of the International Cigarmakers Union. He had established his shop in Anaconda in its beginning years and it was said that he held the second oldest union membership in town. Con McHugh was a blacksmith's helper at the foundry and a member of the executive committee of the Central Labor Council. Michael Tobin was a laborer on the smelter. The elected council members were Elias Jacobson, a Norwegian furniture dealer in the Fourth Ward, Ludwig Adler a saloon keeper in the Fifth ward and Joseph H. Schwend a lumber dealer in Sixth ward.

The Socialists were elected on a platform of municipal improvements and a better deal for the working man. It called for social justice and the immediate public ownership of local utilities, the granting of an eight-hour day to city employees, free public health care, and education reform that included kindergartens, manual training, free textbooks and free medical attention, teaching of economics and history emphasizing child labor laws and the rights of children. The Socialists put on a vigorous and well-organized campaign and their message played particularly well in the east end of town, "Goosetown", which had the largest concentration of new immigrants and day laborers. It was the votes from the Fourth, Fifth and Sixth Wards, all on the east end which gave Frinke, Tobin and McHugh their winning plurality over the other two parties and elected Socialist council members. [5]

The victory came as an unpleasant surprise to the two traditional parties as well as to the ACM. Perhaps because it had been concentrating on what up to then had been the far more serious struggle with Fritz Heinze in Butte, ACM had overlooked its political interests in Anaconda. But the election results were a wake-up call that brought swift and sure punishment for Anaconda smeltermen. The Company began systematically to fire every Socialist or Socialist sympathizer in its employ. By the end of May, the Mill and Smeltermen's Union Local 117, claimed that 150 of its members had been fired. Socialist members of craft unions were suffering similar fates. What were the unions to do? Should they declare every citizen's right to vote his conscience and go out on strike to protest the firings and protect that right?

They first tried to negotiate with Anaconda Company General Manager William Scallon, who stated bluntly that "when a man goes out talking coal oil and Standard Oil and kerosene...can that man compel me to employ him?" For Scallon the matter was closed. No Socialist would be hired and any known Socialist on the payroll would be fired. Getting no satisfaction in that quarter, the union appealed to the Western Federation of Miners for assistance in getting those fired reinstated. The Federation, while sympathetic, pleaded a lack of resources because of its support of active strikes in Cripple Creek, Colorado City and Telluride, Colorado. Nor was help coming from the Miner's Union in Butte, whose leaders were unsympathetic toward socialism. If after these reverses the unions had any thought about striking, they were made vividly aware of their weakness against Amalgamated in October when the New York offices shut down all its operations in Montana as a result of its ongoing fight with Heinze. Now the whole town was out of work. There is no record of the total number of men the Company fired for voting Socialist nor how many may have been taken back in November when the smelter was reopened. But the Company had made its point, just as Marcus Daly had in 1893, it would not tolerate political deviance (Populists or Socialists) in Anaconda. A year later Con

McHugh was reported as saying, "Comrades who had taken a prominent part, or had been heard to express Socialistic sentiments, or who
had been detected trading at Socialist stores, were notified that their
services on 'the hill' were no longer needed. Thus, they got the
radicals out of town and forced the others to cease open activity." [6]

But what about John Frinke, the Socialist mayor, his treasurer
Tobin, police magistrate McHugh and the Socialist council members?
How was the city to deal with them? The City Council had six
holdover Democrat and Republican members, plus two new Democrats and one new Republican from Wards One, Two and Three. The
Socialists were outnumbered three to one on the Council and the
majority took its lead from the Company. First the city clerk invoked
a technicality to stop the new officers from being sworn in and the
six holdover councilmen boycotted the City Council meeting at which
the swearing in was to take place. It was not until June and after a
judicial ruling that the new officers were permitted to take office. But
they were never allowed to govern. The council majority rejected the
mayor's nominees for city jobs, held up the pay of Tobin and
McHugh and generally made life difficult for the new Socialist
incumbents. The only part of the Socialist platform that was enacted
into law was an eight-hour day for city employees. [7]

It was not only the municipal elections of April, 1903 that
sounded alarm bells in the Company offices in the Hennesy building
in Butte, it was something far more threatening that began in the
Deer Lodge county assessor's office at about the same time. Socialist
Peter Levengood, elected the year before, took steps to equalize the
tax burden by closing the loopholes through which millions of dollars
escaped taxation. He hiked the assessed valuation of the Washoe
Smelter from $795,000 to $1,315,000 and increased the assessed
value of property in the county from $7,579,021 to $15,881,178, two
thirds of this was to be absorbed by the Company. The three county
commissioners, all Democrats and beholden to the Company, quickly
rejected Levengood's evaluations and reduced them to those of the
previous year. Thus the Company, still in total control, waged its war
on Anaconda Socialists. According to Matt Kelly, it was at this time
that Phoebe Hearst decided to deed the Hearst Free Library to the
City rather than pay taxes on it. [8]

It was apparent the following year, though Frinke was still
mayor, that the Company had completely eradicated its hated enemy.

In the municipal elections that spring the Socialists were buried in a landslide of votes in all six wards. The *Standard* predicted, "In all likelihood nothing more will ever be heard of the Socialist party in Anaconda."[9] And the *Standard* knew whereof it wrote. The following year, Frinke, Tobin and McHugh all ran for reelection and were defeated receiving only one-quarter of the total vote. Never again was the party of Eugene Debs to rear its threatening countenance in Anaconda.

In Butte the story was different. With the feud between Heinze and Amalgamated at its height, the fight for votes was so critical to both sides and the campaigning so charged that the Socialists hardly made a showing. It was not until after Heinze had left that Socialist strength began to increase, paralleling a similar rise in the rest of the country. In 1904, Debs got 402,000 votes for president, 6000 coming from Montana, mostly out of Butte. It provided the same amount in 1908 and jumped to 10,000 in 1912, when the national vote for Debs increased to 900,000. The year before, Butte had elected the Socialist, Unitarian Minister Reverend Lewis J. Duncan, as well as a Socialist city council, making it one of 32 cities in the country, including Berkeley, California and Flint, Michigan, controlled by the Socialists. In 1912, Duncan ran for Governor, not winning but getting 12,000 votes. Even though Amalgamated took steps to prevent it, Duncan was reelected mayor of Butte in 1913 for two more years.[10]

As the Socialist Party grew in strength and political respectability on a national level, there was a reaction and a counter-current of radicalism in the national labor movement. Both Butte and Anaconda unions were affected by these developments. The Butte Miners Union was the largest and most important union in both the Western Federation of Miners and in the recently formed American Labor Union which had its headquarters in Salt Lake City. They both needed regular contributions from the miners' union dues to be able to operate. Even though the Western Federation was founded in Butte and had its headquarters there until 1899, it had expanded and was now dominated by non-Butte officers.[11]

Butte had enjoyed twenty-three years of labor peace under Daly's baronial concern for his Irish clan. But elsewhere, the Western Federation's Big Bill Haywood and Ed Boyce had been fighting the traditional, confrontational mine managers in some of the West's most bitter strikes in Coeur d'Alene, Salt Lake City, Telluride, Cripple Creek and Idaho Springs. While Daly's methods had produced a stable, productive labor force concerned primarily with steady work, Haywood and Boyce had been organizing the highly mobile, mostly single, rough and ready itinerants of the gold mining camps whose answer to the crude power of mine operators was violent collision.[12]

Typical of the problems they faced was the strike called in Coeur d'Alene in 1899, when a wage dispute, ballooned into a struggle between the Bunker Hill and Sullivan company and the WFM over union recognition. When Bunker Hill employed non-union scabs to replace strikers, union members commandeered a freight train, stopped it at the company's recently completed giant concentrator near Kellogg, and attempted to force the scabs inside to leave. Sporadic shooting broke out and two men were killed, one on each side. The attackers then dynamited the concentrator and burned down the superintendent's house. The mineowners appealed for help to Idaho Governor, Frank Steunenberg, who in turn asked for federal assistance, since the Idaho National Guard was serving at the time in the Philippines. President McKinley sent in two brigades of black troops to restore order under the direction of Idaho State Auditor Bartlett Sinclair. As in 1893, seven hundred union members were rounded up and thrown into miserable bull pens. Sinclair then proceeded to break the WFM in Idaho by exacting promises from bull pen prisoners to resign from the union as the price of freedom. Union leaders were tried and convicted of the two murders even though they had not even been in the neighborhood of the shooting.[13]

From these experiences and their personal backgrounds Haywood and Boyce became vocal advocates of militant industrial unionism and steered the WFM in that direction. Even though the Butte Miners Union had a reputation for being the toughest union in the West, by 1900 the core of its membership preferred accommodation to confrontation. The Anaconda Smeltermen, who up to that time had never been out on strike, were similarly inclined. These divergent currents put a heavy strain on relations between these important locals and their federation.[14]

Then, in 1905, one of the most bizarre episodes in the history of industrial unionism began. The anti-Fabian leaders of the old

Socialist Labor Party decided to form a new industrial union, one that would be militant and openly revolutionary in its objectives. This was the International Workers of the World, the IWW, the notorious Wobblies. The intellectual promoters of this new union, William Trautman, Isaac Cowan, Daniel DeLeon, Ernest Unterman, and Fr. Thomas J. Hagerty were socialist pamphleteers and agitators. The only union base they had were Unterman's German dominated unions, the Brewery Workers in Milwaukee and Chicago. They needed much broader representation to launch their new union. So they convinced Big Bill Haywood, Ed Boyce, Vincent St. John and Charles Moyer to become founding members. As a result, the WFM became the muscle and sinew of the new IWW. At the urging of Haywood and Boyce, representatives from both Butte and Anaconda locals attended the founding convention and became affiliates of the new organization.[15]

The IWW was an open advocate of class warfare and its revolutionary watchword was "Abolish the wage system". It had a grandiose plan to organize the American working class into one gigantic industrial union which would combine with similar unions in other parts of the world. It was to be organized by departments representing all aspects of industrial production which mirrored the American capitalistic system. These departments were meant eventually to replace the capitalistic structure with a workers' commonwealth. By a leap of faith like that which had prompted the Irish societies of Butte and Anaconda to train for the invasion of Canada, militants of the IWW looked forward to a world revolution which would be achieved by a series of strikes that would turn into a worldwide general strike and force capitalism to crumble.[16]

Far from fantasy, the IWW and its leaders were dead serious about their goals. They set out to organize the unskilled foreign-born in the mass production industries. In the East, where the intellectuals of the movement concentrated, they actively recruited the most recent immigrants, Poles, Russians, Serbo-Croats, Italians, Spanish and Portuguese. Like so many of those who had come before them, these new workers were exploited and discriminated against wherever they went. In one typical example an advertisement for laborers to build a reservoir in Croton, N.Y. in 1895 announced that the pay for common labor was $1.30 to $1.50 a day; for colored $1.25 to $1.40; and for Italians $1.15 to $1.25. This was certainly fertile ground for the IWW, which spoke out movingly for social justice on behalf of those who had no voice of their own against unequal treatment. They sought to represent the weakest sectors of labor, the unskilled, the unorganized and the unwanted. In the West, where the Western Federation of Miners reigned supreme, the IWW had its greatest success among single, male, migratory workers in the mining and timber industries.[17]

But the marriage between the IWW and the WFM was tempestuous and short-lived. It was interrupted shortly after it began as a result of the murder on Dec. 30, 1905 in Caldwell, Idaho of Frank Stuenenberg, who had been governor at the time of the Coeur d'Alene strike of 1899. Two of WFM's principal officers, Charles Moyer, president, and Bill Haywood, secretary, were arrested in Colorado, along with Charles Pettibone, who had been a union officer in Coeur d'Alene in 1893. All were secretly spirited to an Idaho jail and there charged with the murder. Their absence from the union scene for the better part of two years greatly diminished WFM's influence in the IWW and increased tensions between the leadership groups.[18]

The trial itself caused a sensation throughout the United States with Idaho Senator William E. Borah leading the prosecution and famous trial lawyer Clarence Darrow, the defense. The prosecution's entire case was based on the confession of a shadowy figure with the unlikely pseudonym of Harry Orchard, who also went under the name of Thomas Hogan, but whose birth certificate bore the name Albert E. Horsley. Orchard, who had been a member of the WFM, confessed that he had been hired by WFM to kill Steunenberg and that he had also attempted to take the life of Governor Peabody of Colorado, as well as those of prominent jurists, mineowners and superintendents in Colorado. In fact, hardly any unexplained mine explosion or dynamiting of mining property in the West was left unaccounted for in his confession. A corroborating confession by one Steve Adams was repudiated and later proved to have been coerced by the prosecution.[19]

The surreptitious arrest, secret jailing and transfer to Idaho of Haywood, Moyer and Pettibone was condemned as kidnapping by the press across the U.S., but it did them little good. When Darrow appealed the arrest, it was upheld by the Supreme Court. Eugene Debs rose to defend the accused with one of his most inspired articles. Under the title of "Arouse Ye Slaves" he proclaimed, "Their only crime is their loyalty to the working class. If they hang Moyer and Haywood they've got to hang me... Moyer, Haywood and their comrades had no more to do with the assassination of Steunenberg than I had." Referring to a prediction by the chief investigator for Idaho that Moyer, Haywood and Pettibone "would not leave the state alive", he threatened, "If they do attempt to murder Moyer, Haywood and their brothers, a million revolutionists, at least, will meet them with guns."[20]

Convinced that the murder and the arrests were a gigantic conspiracy hatched by the Mine Owners Association and the states of Idaho and Colorado against the WFM and its leaders, the union movement in the West was in a high state of agitation. The Butte and Anaconda unions, along with other WFM affiliates, passed resolutions condemning the arrests and voted special assessments to help pay legal fees. The Socialists of Montana, taking Deb's threat to heart, wrote him saying that they could have 10,000 men, each with two horses, Winchesters and 200 rounds of ammunition ready to meet him at the border and rescue the accused from the Boise jail.[21]

There was some reason to believe the conspiracy theory. Mine owners, probably alarmed by IWW rhetoric, were bearing down heavily on union activity in mines and smelters throughout the West, some refusing to hire union members and discharging workers who refused to resign from the union. The Mine Owners Association had permanent contracts with detective agencies whose agents infiltrated labor organizations to gather inside information on union activities and encourage disruptive acts by the unions to discredit them in the eyes of their members and the general public. Many a strike call came from labor spies, working under cover, to maneuver the unions into ill-considered and unwinable confrontations with employers. James McPartland, head of the Denver office of the Pinkerton Detective Agency, was one such operator working for the Mine Owners Association of Colorado in its fight against the WFM. It was not viewed as a coincidence that he became the chief investigator for the state of Idaho in its actions against Moyer and Haywood.[22]

Orchard himself was a shadowy figure who seemed to show up in any location where there was labor trouble. He had been in Coeur d'Alene in 1899 and in Cripple Creek during a strike in 1903. After his confession, the *Miners' Magazine* claimed that he was in

Cripple Creek as a labor spy working for the Scott and Sterling detective Agency. But he was a member of the WFM and had gotten a letter of recommendation from George Pettibone to work for an insurance company.[23]

Since the case against Haywood seemed to be stronger than that against the other two, the state of Idaho decided to try him first. Even though Orchard withstood a withering cross examination from Darrow, there was no evidence to corroborate his testimony. The jury found Haywood not guilty. Pettibone, too, was brought to trial and acquitted, and the charges against Moyer were then dismissed.[24]

While the WFM officers were vindicated, their arrest and trial had taken its toll on the WFM and on the men themselves. Month after month of unfavorable publicity left the impression in many people's minds that WFM was a violent and unsavory organization. Also, Darrow had taken a strong dislike to Haywood during the trial and recommended when it was over that he not continue in his position as secretary of the WFM and that he stay out of the public eye for a long time. In addition, there were strong suspicions in some quarters that Haywood knew more about the Steunenberg murder than he ever told. The trial had also exacerbated strong personal differences between Moyer and Haywood. Moyer returned to the presidency of the Federation but made no effort to bring Haywood back as secretary. Haywood, after a brief period, became active in the IWW.[25]

These happenings in Idaho and Colorado were playing to a very different audience at the local level. By 1905, the majority in the Butte Miners Union and the Anaconda Smeltermen's Union were family men, husbands and fathers, many of them earning enough to own modest homes and have a stake in the community. Besides this, many had been intimidated by Amalgamated's hard ball tactics against the Anaconda Socialists and the statewide shutdown of all the Company's operations in 1903. In Butte there was an additional incentive not to rock the boat. This was the long established practice of "leasing" that offered ambitious, hard-driving miners with connections the opportunity to become capitalists.

Sometimes, when a mining crew struck a particularly promising pocket of ore, they were able to increase their earnings by convincing the company to lease that part of the mine to them. This was a well-established practice in Butte, where companies used it to reward those miners who had been helpful or useful to management. The lease was a typical way of paying-off a union official who had been helpful in labor negotiations. While there were complaints and protests against this practice by the rank and file within the unions, it became part of the fabric of company-union accommodation, helped keep labor peace, and raise the incomes of a few individual miners. Since officers in the BMU were elected for only one year, annual elections served to curb blatant abuses by individual officers and provided an opportunity to change them frequently, thus either keeping them honest or, at least, spreading the wealth.

So, even though working conditions continued to be harsh and dangerous and living standards remained low, the large majority of local union members had a different orientation than many of the leaders in the WFM and the new IWW. Those in Anaconda and Butte were primarily interested in steady work, job security and safety at the work site.

The IWW's openness to itinerant workers, and the group's mindless radicalism, made it threatening to most members of the Miners and Smeltermen's unions. Also, because most of the workers in both cities were at least nominal Catholics, the Church's stand against Socialism influenced many. In addition, established workers were not sympathetic to new immigrants, most of whom were un-skilled, didn't speak English and would work for low wages. They viewed them as "outsiders", a threat to job security and a menace to safety at the work site.[26]

This was particularly true in Butte, where the Irish and the Cornish had long dominated the mines. Though these two groups fought bitterly among themselves, sometimes leading to tragedy as in 1901 when Cornish editor W.J. Penrose was murdered for voting against the eight-hour day in the state legislature. But both were proud of their skills as drillers, trammers and timbermen and resisted letting newcomers into their ranks. While they would gladly train newly arrived members of their own tribes, they didn't welcome working beside "greenhorns and greasers". This put the union membership at odds with the IWW and the Western Federation of Miners. Ironically, it also put them at odds with the new owners of Butte and Anaconda, Amalgamated Copper.[27]

22. Silver Jubilee

After the Heinze buy-out in 1906, the Butte Miners Union, remembering the massive 1903 shut-down, made peace quickly with the now all-powerful Amalgamated Copper Company, signing a new five-year contract. Wages were set according to the price of refined copper for the calendar month. When the price exceeded 18 cents a pound, the pay for work underground was $4.00 a day. Under 18 cents, which was most of the time, it dropped to $3.50 a day. The eight-hour day remained in effect. It was during these negotiations that the Anaconda Smeltermen's Union achieved the eight-hour day for the first time.[1]

The Western Federation of Miners opposed the contract and declared it void. But the local unions, believing that they had preserved both the eight-hour day and the uniform rate underground, were more interested in steady work than in upholding some abstract principle established by the Federation. They refused to back down. In Anaconda the Mill and Smeltermen's Union, acknowledging that work on the surface could not command the same pay as work underground, accepted a $2.75 a day rate for unskilled labor and a sliding scale of $3.00 when the price of copper was above 18 cents a pound.[2]

The union leaders in Butte and Anaconda were not unhappy with the results of their negotiations. They had preserved all their earlier gains and had introduced the possibility of increasing their wages during the contract period if the price of copper improved. In addition, the Anaconda union had finally gotten the eight-hour day for smeltermen. They believed that they had served their membership well and had reflected its wishes. But the WFM, rebuffed by the two local unions, concluded that their leadership had gone soft and set out to remedy the situation. As a result, the union elections of 1907 turned out those who had negotiated the contracts and replaced them with more militant officers more in tune with the WFM leadership, if less responsive to local union membership.[3]

This change was not unrelated to the economic downturn brought on by the Panic of 1907. Thus, the last stages of the Heinze-Amalgamated feud returned to the place of its origins and struck with a vengeance. In December of that year, the mines in Butte and the smelter in Anaconda were shut down and would remain closed for three months. It was a long, cold winter, and a joyless, jobless Christmas for the two communities. And for Anaconda, which had escaped the ravages of the 1893 Panic, it was a grim reminder that the town's health and well-being depended on a single commodity and a single employer.

But 1907 had been a disaster from the beginning. In the early months of the year, there were strikes in various parts of the state by telephone, telegraph and street railway employees as well as by janitors, meat cutters, teamsters, waiters, drug clerks and machinists all demanding higher wages. In Butte and Anaconda there were no newspapers for six weeks as the result of a strike by printers, pressmen and stereotypers.[4]

The freezing winter months of January and February 1908, with no smoke coming out of the big stack, were hard on everyone, causing many to wonder whether the skeptics hadn't been right that Anaconda might become a ghost town, after all. So when the mines and smelter finally reopened on March 8, it was a happy day in both cities. Women made up lunch buckets for their men heartened by the thought that after three months without work they would once again have enough money to put bread on the table. The mood was one of open jubilation.

In this spirit, the city fathers began preparing to celebrate the city's silver anniversary. They were anxious to demonstrate not only that the town had survived the recent shutdown but that its first twenty-five years were merely the prelude to a brilliant future. It would be a vibrant part of a growing America that had already surpassed all the countries of Europe to become the world's leading industrial and manufacturing power.

On the occasion of this silver jubilee celebration, Charles Eggleston wrote, "Twenty-five years ago there were those who fancied that Anaconda lacked the elements of permanency and would not long endure...Instead of verifying these gloomy forecasts, the city, barring two or three temporary interruptions, has steadily progressed until it ranks as one of the most prosperous communities of the

West...''[5] A giant parade on the Fourth of July climaxed the anniversary festivities and announced to the world the town's renewed confidence.

This spirit of progress and optimism continued into the following year when Anaconda again made extravagant preparations, this time for the visit of the President of the United States, William Howard Taft, on September 27, 1909. With Mickey Campbell of Anaconda at the controls of BA&P engine number 20, the Taft party was whisked from Butte to Anaconda in 36 minutes. When Taft's train arrived at the smelter at seven in the morning, he was greeted by 300 workers coming off shift. After a quick tour of the smelter, the President was ushered into E.P. Mathewson's decorated car and, escorted by 12 mounted forest rangers, was driven from the smelter to town. Before arriving at the city common for the official ceremonies, the car made a quick detour to 105 East 7th Street, the home of Dr. and Mrs. S.T. Orton. Mrs. Orton was the President's niece. At the common, Mathewson, head of the reception committee, introduced the President to the waiting crowd, where he delivered a short speech. After the speech the President was escorted to the BA&P station, where he boarded his train for Butte.[6]

In the process of preparations for the presidential visit a bitter rivalry developed between the Mayor, Fred Gagner and Mathewson over who was to be in charge. Gagner insisted that as Mayor, he would extend official greetings to the President and complained loudly that Mathewson wasn't even an American citizen. But Mathewson, of course, prevailed. If there was ever any doubt about who was the dominant figure in the city, this little tiff was a fresh reminder.[7]

The 1907 shutdown, while unsettling, had not stopped continued growth and investment in Anaconda. Once production was resumed there was further expansion of the new smelter and the railroad yards were enlarged to accommodate increasing tonnage out of Butte. In a mood of civic betterment, the city council had passed resolutions the year before that all the wooden sidewalks should be replaced by concrete, that cesspools within the city were a health hazard and all houses must be connected to the sewer lines. Much of this had already been done giving the town a neater, tidier look.[8]

Anaconda was also beginning to feel the effects of a gradual shift in patterns of immigration to the United States which had begun before the turn of the century. Industrialization and an improving economic conditions in Northern and Western Europe had slowed the outflow of Irish, English, Germans and Scandinavians. Extreme poverty, political turmoil and natural cataclysms caused a flood of migrants to the United States from Italy, Hungary, Bohemia, Slovakia, Slovenia, Croatia, Serbia, Greece, Poland and Russia. Also there was a dramatic increase in the number of new immigrants. For example, from 1871 to 1880 new arrivals totaled 2.5 million, most of them from the British Isles and northern Europe; from 1891 to 1910 the total was 12.5 million, seventy percent from southern Europe. Between 1881 and 1890 alone, one million immigrants arrived from southern Europe. From 1891 to 1900 this number doubled to 2 million. Then from 1901 to 1910 it tripled to 6 million.[9]

The United States was rapidly expanding its industrial base and mine and factory owners welcomed the increasing flow of low-wage labor. Most new arrivals stayed in the large industrial cities of the East or were absorbed into the steel industry and the coal mines. But Amalgamated was able to attract considerable numbers to continue on west to work in the mines and smelters of Butte and Anaconda. As earlier immigrants came through the Port of New York tagged for Anaconda in care of Superintendents Dougherty, Kelly, Dalton or O'Brian to work for Marcus Daly, they now came to work for Amalgamated Copper.[10]

The new immigration trends were also reflected in the workers being brought in to work on the BA&P railroad. Italians, most of them direct from southern Italy, working under Irish bosses, made up many of the railroad section gangs laying tracks, and added to the already diverse ethnic mix of the town. They did not join the small community of earlier Italian immigrants, who were mostly from Tuscany, but settled instead in a small enclave on west Commercial street not far from the main office of BA&P railroad. On the smelter, the number of Croatians and Slovenes continued to grow, now joined by Serbs and Montenegrins. Known collectively as Austrians, most of them settled in the new eastern addition. It was during these years, also, that the number of blacks working on the smelter increased slightly to about twenty-five, and the African Methodist Episcopal Church, which had functioned since 1895, moved the old school house from Carroll to west Commercial Street and converted it to a church.

There was also, for a brief time, a Japanese presence in the town. A Japanese section crew was brought in to lay new track for the Mill Creek spur line to the new smelter. This line, later called the "loop", over a steep and winding grade, was needed to get the ore to the top of the hill above the new concentrators. The Japanese workers lived in bunk cars while the job proceeded but, unlike the Italians, departed once the job was finished.[11]

The area around Mill Creek became especially active during those years. The timber activity out of French Gulch continued. In 1911, the Anaconda Company terminated its contract with the Allen Timber Company and formed its own Mines Timber Company to continue the work, with Nels Pearson as general manager. Nels was born in Skane, Sweden and had come to Montana in 1886. At the time Anaconda hired him, he was running his own timber operation in the mountains south of Gregson hot springs on Willow Creek. The Mines Timber Company had two camps, one at French Gulch and one at Mill Creek. In addition to the camp, Mill Creek also boasted a hotel called the Ten Mile House, a boarding house, a store and a few cabins on the road near the top of the pass. This was a large operation with hundreds of men and even more horses and mules. There were loggers, saw mill operators, teamsters, mule skinners, and a couple of veterinarians and bookkeepers and clerks. In each of the camps there were blacksmith shops, stables, saw mills and commissaries. Along the fifteen-mile wooden flume used for floating the sawed logs to the railhead were watchmen's cabins where men lived with their wives and family and watched the flume for log jams.[12]

Once the "loop" to the smelter was completed, the flume, which carried the logs to railroad sidings near the BA&P mainline, was shortened to connect with the new railroad spur. Over twenty railroad cars a day were loaded with timber and hauled to Butte to provide props and stulls for the mines. By this time, the smelter was no longer using wood in its furnaces, but had converted to coal and coke. This activity out of Mill Creek continued until the end of World War I, by which time the area had been virtually logged out, and what little timber that was left had died from smoke damage. Operations, still under Nels Pearson's supervision, were then moved to Georgetown Lake where large-scale timber cutting continued until 1926.[13]

11. View of town and smelter looking southeast. Washoe Park in foreground. About 1924.

During these same years, a completely new enterprise, un-related to the smelter, grew up around Mill Creek due to the activities of the Montana Meat Company, founded by David D. Walker and Nicolas Beilenberg. After the BA&P railroad was constructed they built a new slaughterhouse next to the railroad where it crossed Mill Creek. From the time the slaughterhouse first opened it had attracted the Indians under Chief Rocky Boy who soon came to rely on hand-outs of meat scraps, innards, horns, hooves, damaged hides and other leftovers. They camped along the creek and hung around the slaughter house.[14]

The area became even more lively once feeding pens were built to hold cattle in transit. As early as 1883, Beilenberg established a ranch in the Big Hole valley and began bringing in cattle every fall for winter feeding and spring sale in Butte. In the summer, hay was made from the abundant and nutritive native grasses, then cut and stacked for winter feeding. About 1900, as the reputation of this grass-fed Big Hole beef spread, buyers from packing houses in Seattle, Tacoma and Portland began making purchase contracts and Big Hole ranchers expanded their feeding activity. By 1905, David Walker had two ranches in the Big Hole of over 10,000 acres, pro-ducing almost 7,000 tons of hay. Cattle for winter feeding were trailed from the railroad stock yards on Mill Creek and over the divide into the Big Hole. In the spring they were trailed back out for shipment to the West Coast, some being transshipped as far away as Valdez and the Alaskan gold fields. Once others got into the business, cattle were shipped east to Chicago, as well. During peak years, fifteen to twenty thousand head of cattle a year, mostly from Texas, made this trek. In 1914, 30,000 cattle were shipped out of the Big Hole to market. The *Anaconda Standard,* waxed enthusiastic when State Cattle Inspector John Collins stated that Anaconda had become the largest cattle shipping point in the state. It quoted him as saying that the famous hay of the Big Hole basin was the best cattle feed in the world and that Big Hole beef as good as the best corn-fed product from the Middle West. Such praise combined with the complete absence of disease made Big Hole cattle the best in the world. The article also pointed out that as the older native grass pastures were replaced by clover and timothy, the hay lost none of the quality that had made the basin a stockman's paradise. It appeared that Anaconda was about to have a second major industry. But after World War I and into the 1920s, the business gradually declined. After Walker died in 1906, the slaughterhouse was taken over by W.W. Montgomery. He continued to operate even after large-scale cattle feeding in the Big Hole had tapered off. By then, grass-fed Big Hole beef had established a lasting reputation on the local market and continued to command premium prices.[15]

All of this activity was especially welcome by the owners of the Three Mile House at Mill Creek. It had originally opened to slake the thirst of lumberjacks and railroaders and then became a favorite haunt for cowboys, cattle buyers and shippers, as well. During the summers, horse auctions were held there every Sunday. Since most of the horses were wild and unbroken, buyers would often pay to have them ridden, making bronco riding and bucking contests another Sunday afternoon attraction for townspeople.[16]

Even with these increases in economic activity, however, the overall population of the county was down from the high it had reached during the construction of the new smelter. The census for 1910, reflecting the splitting off of Powell County, indicates that the population of Deer Lodge County had declined from 17,393 in 1900 to 12,988.

23. John D. Ryan

These years saw the start of a new run of economic prosperity for both communities. Butte reigned supreme among mining cities of the West and the future for copper looked ever brighter as the electric light bulb was replacing gas light and the kerosene lamp. Electrification became a high priority in American cities, and Europe increased the importation of American-made copper products. John D. Ryan, out of Lake Superior copper country and president of Marcus Daly's bank in Butte, had made a good impression on Amalgamated's Henry Rogers and had taken over Daly's role as president of the Anaconda Company. It was Ryan who negotiated with "Fritz" Heinze and put together a satisfactory purchase arrangement for his Butte properties. Together he and Rogers directed the continued consolidation of assets on Butte hill and the rationalization of production practices. They pulled together the various companies operating Amalgamated's more than one million acres of woodland with mills in the Bitter Root, Bonner and St. Regis, and its coal mines and water resources across the state into an integrated system. All of the Amalgamated properties, including the Heinze mines and most of Clark's, were now operated under a single management structure, with headquarters in New York City.[1]

And there was no doubt that the Anaconda Company would continue to expand and with it the operations of the Washoe Smelter. When Ryan took over as president of Amalgamated Copper after Henry Rogers died in 1909, he began to phase out Amalgamated as a holding company transferring its assets to the Anaconda Company. At the same time he continued to consolidate and integrate all phases of the production process. In 1911, Ryan and Cornelius Kelley, now an Anaconda vice president, decided that it made no sense to ship the 5000 tons of ore coming daily out of the Boston and Montana mines one hundred and sixty miles to Great Falls for smelting, when this ore could go to Anaconda just twenty-six miles away. Great Falls, with its abundance of cheap hydroelectric power, would become instead the refining center. This meant that the Washoe smelter would again increase its capacity and work force to handle the additional ore. This was good news for the town and a sign that if the price of copper held, there'd be plenty of work for the foreseeable future.[2]

Early in 1910 all of Amalgamated's operating companies were folded into a single corporate entity. Thus, the consolidated Boston companies, the Trenton Mining Company, the Big Blackfoot Lumber Company, the Red Metal Mining Company, the Parrot, the Alice, the Colorado, and the Washoe Copper Company, which operated the smelter, all became part of the new Anaconda Copper Mining Company, thereby increasing its capital stock value from $30 million to $150 million. The Clark properties, the Original and Stewart mines and the Butte Reduction Works, bought by Amalgamated were merged with Anaconda in May 1910 and the Heinze properties in June 1911. Amalgamated was now only a shell and Anaconda emerged as a completely integrated New York-based world-class corporation ready to do business not only in Montana but wherever the opportunity presented itself.[3]

The economic activity generated by Amalgamated's investments in Butte and Anaconda attracted the interest of a new large enterprise, namely another railroad. When the Anaconda Company was taken over by Amalgamated, William Rockefeller's financial interest in The Chicago, Milwaukee, St. Paul, and Pacific Railroad forced itself onto the Montana stage. Daly's long alliance with James Hill and the Great Northern was soon superseded by Rockefeller's ties to the Milwaukee. The Milwaukee bought out the Great Northern's investment in the BA&P, and began to construct a western extension of its own road in 1906. It entered Montana in 1908 and reached Gold Creek in 1909 following the same route from Butte to Deer Lodge as the old Montana Union railroad, utilizing the BA&P right-of-way through Silver Bow Canyon, to Stuart. Butte, which was already served by the Union Pacific and the Northern Pacific (each with four trains daily) and by the Great Northern (with two daily trains) was now connected to the rest of the United States by four major railroads, confirming its importance as a mining center. Prospects for the continuing growth of the area appeared bright, indeed.[4]

However, even though Amalgamated had achieved total control of copper production in Montana, it never reached the goal which Rogers and his associates had originally set for it of monopolizing and price-setting for the entire U.S. copper industry. The prolonged fight with Heinze had distracted it and delayed its time-table while competitors in other parts of the country made new discoveries and production breakthroughs. Since 1875, with the discovery in Bisbee of the Copper Queen mine, Arizona had been an important producer of copper. Further developments in Globe, Jerome, Douglas, Miami and Ajo, moved Arizona to the front ranks, out-producing Montana. Then Daniel C. Jackling and his associates opened up Bingham Canyon in Utah to open-pit mining which made available huge quantities of lower grade ores pushing Utah into second place after Arizona. Even though total demand for copper was growing rapidly, this exponential increase in production created very stiff price competition. If Anaconda were to continue as an industry leader, it had to find dramatic ways of reducing costs.[5]

Ryan believed this could be achieved through the extensive use of cheap hydroelectric power. Electricity had played an important part in copper production processes in both Butte and Anaconda from the start. The old Upper Works used steam generators at first and later hydro power from a local power plant. In 1906, the Company purchased the power house at Flint Creek which had originally supplied electricity to the town of Philipsburg and the Granite silver mines. At that time, it replaced the steam pumps at Silver Lake with electric pumps and brought power from Flint Creek to the Washoe Smelter power house so it could add additional roasters, a larger blast furnace and two new stands of converters to the smelter.[6]

The Butte mines used steam generators and later tapped into hydro power coming from the Big Hole. As production increased, so did the use of electricity. Amalgamated began in 1907 to replace the horses and mules with electric locomotives to haul the ore in the mines. But the unit cost of electricity was high. Also, as the mines got deeper, some already a half mile into the earth, problems with ventilation and accumulation of water at lower levels got steadily worse. Steam-powered electric generators, pumps and fans used in individual mines became increasingly expensive to operate and added considerably to production costs. Furthermore, the development of new metallurgical processes made the need for inexpensive electricity even more urgent.[7]

The Company, of course, was not without resources. When Amalgamated took over the consolidated Boston companies, it acquired a part interest in the power plants at Black Eagle Falls on the

Missouri, which generated 140,000 horsepower for the Boston companies' smelter near Great Falls. In 1901 Amalgamated also helped finance the Missouri River Power Plant at Canyon Ferry near Helena, and invested in the Butte Electric and Power Company plant in the Big Hole.[8]

Even before Ryan took over as head of the Anaconda Company, he had been interested in power development. With commitments from Anaconda to buy the power, he had formed the Great Falls Power Company in 1908, built the Rainbow power plant and erected long distance, high-tension transmission lines to Butte and Anaconda, which was a major innovation for the time. After several years of maneuvering, he was able to convince Butte Electric to merge with his Great Falls Power Company and form the Montana Power Company in 1912. Montana Power then bought up the small plants of the Madison River Company, Billings and Eastern Montana Power and Thompson Falls Power. Thus with Ryan heading both Montana Power and the ACM, the Anaconda enterprises would be provided with cheap electricity through control and operation of most of the state's hydroelectric power sources.[9]

Since the power plants in Great Falls and Helena were so far from the major consumers in Butte and Anaconda, Ryan's construction in 1910 of gigantic steel towers to carry high voltage power lines, was truly an audacious and visionary move. The immediate necessity for long-distance transmission of electricity forced him to become a pioneer in the field. His decision was fully justified soon after the lines were completed. He estimated that in the first full operating year after installation, Anaconda had saved between $1.3 to $2 million.[10]

But this was not only a technological achievement of major importance. It was another giant step in Anaconda's domination of the economic and political life of the state. John Ryan was president of both companies (lumped together in popular parlance as The Company) and remained a guiding force in both of them for the next twenty years, until his death in 1933. He had surpassed all his predecessors in personal power and was without a doubt the most powerful man in the state. Unlike Daly, Clark or Hauser, he was unrivaled. No elected official had even a fraction of his power. But it was a power not wielded from within the state, but from afar in New York City. This may have made it more judicious but also more

12. Two BA&P electric engines pulling eighty cars each loaded with 75 tons of ore, shortly after railroad was electrified in 1913. Note Northern Pacific and Milwaukee railroad tracks converging on same route through the Silver Bow Canyon.

impersonal and therefore less trusted by the citizenry. The cynicism and suspicion Heinze planted among miners and smeltermen in his fight with Amalgamated grew more pronounced as Anaconda's power increased. Daly's personal touch, which had generated mutual tolerance and understanding for so long, was gone. Though Ryan, as president of the Daly Bank and Trust Company, was a familiar figure in Butte and Anaconda before he went off to New York, his company, with its Rockefeller connections, was now viewed with suspicion.

One indication of eroding labor-management trust came in 1909 when switchmen of the Great Northern Railroad went out on strike. Ryan tried to force miners and smeltermen to work as switchmen until the strike was settled. Instead, the Butte Miners' Union and the Anaconda Mill and Smeltermen's Union voted in sympathy with the strikers. Ryan threatened to shutdown all operations on January 1, 1910 for six months. Again, the unions voted unanimously not to break the strike. Shortly thereafter, the shutdown order was quietly rescinded and work proceeded normally. The Great Northern would have to deal directly with its strikers. Ryan's bluff only increased labor's lack of trust in The Company.[11]

Within a year after the formation of Montana Power, Ryan pushed ahead to expand the use of his new cheap power resource. Since 1910, the company had considered the possibility of substituting steam with electric locomotives on the BA&P run from Butte to Anaconda. In joint studies with the General Electric Corporation it was determined that electrification would reduce operating and maintenance costs as well as increase train size and double mainline speeds. The railroad contracted with GE to build fifteen electric locomotives for hauling ore and two for passenger service. The freight locomotives were geared to pull at least seventy-five fully loaded ore cars and to attain speeds of up to thirty-five miles an hour. The passenger locomotives were geared for forty-five miles an hour. After high voltage lines were brought in from Great Falls, substations were built in Butte and Anaconda to transmit electricity to overhead lines along the tracks. The first electric passenger train departed Anaconda for Butte on October 2, 1913. Ore trains began service the following day. The savings in operating costs were such that the entire project paid for itself in five years. Even before then four additional locomotives and two "half-locomotives", called tractor trucks, were ordered.[12]

The BA&P was the first railroad in the United States to use this many electric locomotives for such heavy hauling, a pioneering effort that was widely studied, both in the United States and in Europe. An important cost-saving feature that it demonstrated was that electric engines pulling heavy trains over steep mountain grades would return almost as much electric power to the line as they used. By using the engine for breaking on the down grades, electricity was generated rather than consumed.[13]

Even before the BA&P project was completed, Ryan, who was on the board of directors of the Milwaukee railroad, convinced the railroad to electrify the Rocky Mountain portion of its line from Harlowtown, Montana to Avery, Idaho. This was a double coup for Ryan. Not only did he get a long-term contract for Montana Power to supply the electricity, but the Anaconda company sold the railroad $5 million in copper wire and electrical supplies for the project. At the time it was completed, this was the longest stretch of electrified rail line in the United States.[14]

24. Expansion and Opportunity

The profusion of technological developments in the latter part of the nineteenth century was by no means limited to the United States. While much pioneering work in the use of electric power was done in the U.S., and the Anaconda Company made an important contribution, Europe led the way in the perfecting the gasoline-powered internal-combustion engine. George Selden of Rochester, New York experimented with gasoline cars as early as 1879, but the work done in Germany and France in the 1880's eventually had as profound an effect on the world as Thomas Edison's inventions were already having. The automobile and the airplane were direct results of this experimentation.[1]

In 1893, Henry Ford introduced a gasoline powered engine to the United States. On Thanksgiving Day 1895 two brothers named Duryea, bicycle mechanics from Springfield Massachusetts, won an automobile race sponsored by the *Chicago Times-Herald* in a car with a gasoline engine. The race caught the interest and enthusiasm of the country and shortly thereafter carriage makers, wagon builders and bicycle manufacturers began to produce gasoline-engine automobiles in quantity. By 1899, there were thirty automobile manufacturers in the United States who in that year alone produced over 2,000 vehicles. In 1908, five hundred and fifteen automobile companies were in business and by 1910, over 450,000 automobiles had been registered. Men like Henry Ford, William Durant and Ransom Olds were already thinking about ways to lower production costs and thus bring down the price of the final product by dramatically increasing volume. By 1906, Ford was producing his Model T for $600.[2]

In contrast, the aviation industry got off to a slow start in the United States, even though it was in this country that the first flight of a fixed-wing aircraft took place. The pioneers of manned flight (again two brothers) bicycle mechanics Orville and Wilbur Wright from Dayton, Ohio, were less successful than the Duryeas had been at igniting public imagination and inspiring the production of airplanes on a large scale. European governments quickly saw the military potential of these new machines and began subsidizing their production. But only in 1908, five years after the Wright brothers first flight at Kitty Hawk, North Carolina, did the U.S. government give the Wrights their first contract, $25,000 for one military plane; and it was World War I before the airplane became more than a curiosity. Only after the war, in 1918, was the first airmail service inaugurated between Washington and New York City.

In Montana, where most roads were still rough wagon trails and even the high-axled Model T had difficult going, the automobile was viewed more as a novelty than a serious rival to the horse. Nevertheless, by 1906 a number of eight-passenger tourist cars and smaller horseless carriages, called gas buggies, began appearing on Anaconda streets. Purchasers of these vehicles had a dizzying array of choices from among such names as Stanley Steamer, Pope Hartford, Pierce Arrow, Winton, Cadillac, Stevens Duryea, Premier, Franklin, Durant, Hudson, Packard, Oakland, Chalmers, Peerless, Hupmobile, Briscoe, Chandler, Mercer, Reo, Diamond T, Studebaker and Oldsmobile. Because of the cold winters, the air-cooled Franklin which had no radiator was an early favorite.

The first automobile dealership in Anaconda was established in 1905, by Charles Branscombe. A native of Massachusetts, he had arrived in Anaconda in 1900 from California and opened a bicycle shop which he operated for five years before getting the Ford agency. His first sale was to Emil F. Ulrich, a plumber, who bought a two cylinder model. According to Branscombe, the first car in Anaconda was a Stanley Steamer bought in 1902 or '03 by Dr. Daniel G. McDonald, a dentist. In 1908, Branscombe moved his dealership to expanded showrooms on the first block of West Park St., where it operated for the next thirty years.

One of the first gas buggies in Butte was a two-cylinder, link-motion gas burner owned by Harry Cole. Cole had won the car in a raffle in 1902 for subscribing to the *San Francisco Examiner*. By 1905, there were 120 automobiles in Montana and the Anaconda Standard reported in 1906 that hunting jack rabbits with an automobile was the latest thing in the way of a summer pastime. The story out of Great Falls related how a 45 horsepower Packard reached sixty miles an hour chasing a jack rabbit for six miles across the open range dodging badger and gophers holes as it went. There were also already reports of wrecks and crashes, such as: "Butte man killed in auto accident in Los Angeles in collision between an electric car and a large touring car."[3]

And it was only gradually that the gasoline engine dominated the market. In 1906, the big news in Butte was the arrival of three steam powered cars and the use of a new electric car for sight seeing. In 1908, the Butte fire department bought a new horse-drawn steam-powered pump engine. Anaconda, eager to keep up with the times and maintain its self-image as a trend setter, acquired a gasoline-powered fire engine the same year. But it would take another twenty years before the horseless carriage replaced the horse on Anaconda streets, before livery stables and blacksmith shops were outnumbered by gas stations and automobile repair shops. As late as the early 1930s a few horse-drawn milk wagons and ice wagons still delivered door to door and a couple of livery stables were still in business.

The Company, converted gradually to internal-combustion locomotion but continued using small electric engines in the mines which had replaced the mules hauling ore from the mine face to the shafts. In Anaconda, smelter superintendent E.P. Mathewson had a gasoline-engine Stevens Duryea eight passenger touring car for his personal use and introduced large, hard-tired trucks with chain drives for hauling materials in and out of the smelter. Some of these trucks were still in use at the foundry well into the 1930s. In the meantime, the BA&P, which had increased the number of daily passenger trains between Butte and Anaconda to six each way in the summer of 1906, expanded its inventory of electric locomotives, and the electric street car line remained the primary link between town and smelter.[4]

But the town was not immune to the fascination of these marvels of the new century. Many took the train to Butte on July 4, 1912, to see Barney Oldfield, known as the world's greatest auto racer. He had been the king of auto racing for thirteen years and was competing against Lewis Heineman, the Vanderbuilt cup winner. They also read with interest the story in the *Standard* about the first flight in Montana by Bud Mars at the State Fair in Helena, September 26, 1910, but were disappointed by the airshow in Butte in 1912 when former Butte resident, Terah T. Maroney, piloting a Curtis biplane never got over seventy feet off the ground.[5]

Nevertheless, the arrival of the first airplane in Anaconda later that year was greeted with wonder and amazement when it set down near the race track west of town and taxied up to the grandstand. It looked like a box kite with a motor and a short tail, not much changed from the machine that the Wright brothers had flown at Kitty Hawk. The pilot was one of a growing number of cross-country barnstormers moving from town to town earning his way by charging the locals for brief flights over their town. He had flown in from Butte and was on his way west. Word came a week later that the pilot had been killed in Spokane, Washington, when his plane crashed.[6]

Flying was still a risky business reserved for the young and the adventurous. The year before, Lincoln P. Rogers was the first person to fly across the United States. The trip from New York City to Long Beach California, had taken forty-nine days, and his plane had crashed nineteen times. The traveling public would continue to rely on trains for the foreseeable future.

Such practical problems didn't phase some of the Butte Irish though, who busied themselves raising funds to be sent to Ireland to build flying machines for a coming war against the British oppressor. There was even giddy talk about dynamite-savvy Butte miners riding the planes over London and dropping dynamite sticks down the chimneys of the parliament buildings.

However, the most important transportation development for Anaconda's everyday existence had nothing to do with airplanes but with the railroad. In 1911, the BA&P extended its line west to Georgetown lake and Southern Cross. The Company had bought the Southern Cross gold mine and trains began to run there regularly hauling out ore and loading timber for the Butte mines. When World War I began, this activity increased and Southern Cross was converted, once again, into a thriving gold camp. This extension of the railroad also proved to be a popular amenity for the town and on weekends families with picnic baskets and fishermen bound for Georgetown lake filled the train.

There was a new transportation link east of town, as well. In 1915, the street-car line was extended five miles to the growing village of Opportunity. Its single motor car dubbed the "galloping goose" could be seen skipping along the tracks four or five times a day on the way to or returning from this new community. The area directly east of the city, which had earlier been farm land, was becoming a vast industrial dump for the residues of the smelting process. The

13. The Washoe Smelter at night, riding the hill above the town like a giant ocean liner.

black slag waste from the furnaces was piled in an ever-growing mound where the old T. C. Davidson house used to sit, between the converters and the Butte highway. And on the valley floor, where Indians once dug bitterroot for their winter stores, streams of strong smelling, chemically polluted waters from the concentrators were run into toxic settling ponds, called slum ponds, which came to occupy over one thousand acres of land. After the pollutants in suspension settled out, the water drained into the Deer Lodge river.

In its continuing campaign to demonstrate that smelter smoke was not poisoning vegetation and livestock in the east valley, the Company formed a fully owned subsidiary called the Deer Lodge Valley Farms Company, headed by Canadian-born veterinarian Henry Gardiner. In addition to cultivating hay and grain and raising sheep, it created the village of Opportunity in 1913, two miles east of the ponds. The Farms Company subdivided land which the Company had previously bought from protesting farmers and offered it for sale in ten acre tracts to Company workers. Irrigation water was made available, at a fee, from both Mill Creek and Willow Creek.

The land sale was meant to give workers who wanted it an opportunity to raise their own pigs, chickens, cows, sheep and horses and to grow their own fruits, vegetables and forage crops, to supplement their incomes from the smelter. This had particular appeal to the new immigrants from rural southern Europe. Soon after the street-car line was completed, there were about fifty families residing there. The Company also built an eight-room school house for students through the eighth grade.[7]

The "galloping goose" was the workers' connection with their jobs and Opportunity's connection with the rest of the world. A few years later, the Anaconda Country Club, formed by Company officials, built an eighteen-hole golf course next to the village. This provided yet another contingent of riders for the new street-car line and jobs as caddies to the youth of Opportunity.

By this time Frederick Laist, who began working in Anaconda in 1904, had become general superintendent. Many of the innovations at the new smelter had come from his fertile mind when he worked in and later headed the research department there. He was the one who had advised Kelley in 1911 to begin shipping the copper ores from the Boston and Montana mines at Butte to Anaconda rather than Great Falls for smelting.[8]

Under his guidance the laboratory developed a process for producing pure acid from ores containing sulfuric acid. He had also set up a pilot plant to test the recovery of alumina from tailings and was experimenting with the production of high-grade phosphate for fertilizer. In 1913, a plant to process tailings with a capacity of 2,500 tons a day was constructed. The following year, the research department undertook a program to develop a process for recovering zinc and other trace metals from the complex Butte ores. This resulted in the first production of electrolytic zinc on a commercial scale in the United States. Plants were built at Great Falls and at Anaconda and zinc production became a major activity at the Washoe Reduction Works.

The electrolytic process made it possible to recover zinc from otherwise valueless ores not only from Butte hill but throughout the world. It also added significantly to the mineral reserves of Butte. The new zinc plant at Anaconda almost doubled the size of the smelter with concentrating, roasting, leaching, electrolytic and casting departments. However, because of increasing gains in productivity the size of the work force was only modestly augmented. Additional electrolytic operations were established at Great Falls, to take advantage of its abundance of hydroelectric power.

These were not the only innovations to come out of Laist's research department. He and his researchers created a process for combining sulfuric acid produced at the smelter and phosphate rock to manufacture phosphoric acid. This acid was then used to treat phosphate ores in the production of superphosphate fertilizer. The new product had triple the potency of commercial phosphate fertilizers then on the market. It was given the name treble superphosphate. In 1919, the Company purchased 3,600 acres of phosphate deposits and a number of mining claims near Soda Springs, Idaho and others at Garrison, Montana. It then constructed a plant at Anaconda to produce the fertilizer, further diversifying production there.[9]

It was also during these years that the Company introduced another innovation in the processing of its raw ore. The Anaconda smelter became the first plant in the world to use the flotation process to concentrate ore. This process, which had been discovered some years before in Colorado but had never been tried in a large-scale commercial operation, consists of making the heavier minerals float in a mixture of water, oil and chemicals and letting the lighter particles sink to the bottom. The finely ground ore is agitated in the water and the heavier minerals become coated with the oil and chemical mixture and float in a froth on top which is skimmed off, dried and subsequently smelted. The proportions of oil and chemicals used determines which minerals will be recovered. The introduction of flotation at the Anaconda smelter not only increased the percentage of mineral in the concentrate but increased the smelter's overall output. It was the run-off from this process that found its way into the slum ponds. As a result of the Company's success, the flotation process became much imitated in plants throughout the world.[10]

These were also the years when the Anaconda Company began expanding its operations well beyond the State of Montana. In April 1914, it bought all the assets of the International Smelting and Refining Company which owned the Raritan Copper Works at Perth Amboy, New Jersey, the Raritan Terminal and Transportation Company, the New Jersey Storage & Warehouse Company, smelters at Tooele, Utah and Miami, Arizona, the Tooele Valley Railroad and the International Lead Refinery at East Chicago, Indiana. These acquisitions gave the company a larger production base, improved its manufacturing potential, and extended its influence into three additional states. In June 1915, Amalgamated Copper, which had been only a shell since Anaconda was reorganized in 1910, was formally dissolved. Ryan then took over Anaconda as president, with Benjamin Thayer and Cornelius Kelley as vice-presidents.[11]

25. Wreck the House of the Grafters

While Ryan and his company were prospering, the union movement in Butte and Anaconda was faltering. This was due not only to the Company's corporate approach to labor relations, but to internal bickering and growing differences between the WFM and the IWW. Ever since the Idaho trial, Moyer began moving the Federation away from the revolutionary unionism and class warfare advocated by the IWW. Then in 1908, WFM broke with the IWW, and Moyer established a working relationship with the United Mine Workers of America, the largest miners' organization in the U.S., which was affiliated with the American Federation of Labor. In 1911, the WFM, with UMWA support, renewed its own affiliation with the AFL. When it came time in 1912 to negotiate a new three-year contract with the Company, the Butte Miners Union itself had once again come under more conservative leadership.

Nevertheless, lingering IWW influence in the WFM and the BMU was to plague both for a long time to come. Evidence was mounting that on some level within the IWW there was a conscious plot to destroy the WFM and local unions which didn't share IWW philosophy. In fact, C.H. MacKinnon, a long-time member of the IWW and Big Bill Haywood's brother-in-law, admitted as much in a statement he made at a meeting of the International Union of Mine, Mill and Smelter Workers in Salt Lake City, explaining his own activities in IWW:

> "My work as an IWW representative is to visit the different mining camps, and as a member of the International, visit the local union office and its meetings and report as to how we can best carry on our campaign. We do not expect to establish permanent unions of the IWW in the camps at this time, but through creating of dissension among the members of the International we will destroy their locals and later when there is no organization...we will go in and organize...and later affiliate it with the IWW."[1]

Whatever the case, IWW planted agents within both WFM and the BMU in much the same way as the Mine Owners Association did. In fact, it became widely suspected that the detective agencies employed by the mine owners used IWW agents to do their dirty work. While representatives from the IWW's national office openly identified themselves, at the local level its agents always denied their membership and swore allegiance to WFM or the BMU.[2]

In 1912, the BMU signed a new contract with the Anaconda Company relating wages to the price of copper. When the price was fifteen cents or less, wages were set at $3.50 a day. When the price reached eighteen cents or more, wages went up to $4.25 a day. Later that year the company instituted a "rustling card" system without negotiations with the union, but which the union officers accepted after-the-fact, even though a majority of members had voted against it. The rustling card was really a way of centralizing the employment and discharge of miners. Up to that time hiring was done at the mine gate before every shift by the mine foreman or superintendent. Now the Company would require those seeking employment to go to a central personnel office, fill out an application listing name, age, nationality and address and be issued a rustling card which entitled the holder to then go to the mine gate to be hired if there were a vacancy. If a miner was discharged he would have to get a new rustling card to be rehired. There was resistance to this system from some quarters within the union because of its potential for Company black listing. The Company claimed it needed the system to comply with both state and federal statutes. It also claimed that it had been in effect at the Washoe Smelter in Anaconda since 1903 and had not been used against union members. The union finally accepted the Company's action, rationalizing that it would also be useful to track miners who had not yet joined the union and to collect dues.[3]

Nevertheless, there were militants within the union, headed by Thomas Campbell, who remained fiercely opposed to the union leadership, the new contract and acceptance of the rustling card. Even though they did not have enough votes to be elected as delegates, they vowed to make the rustling card an issue at the coming WFM convention and maneuvered to be heard there and present charges against BMU officers, Charles Moyer and the WFM leadership for not taking remedial action. Campbell charged that Dennis Murphy, BMU president, and other union officers were company stooges, that the majority of the union was against the rustling card system and that Moyer had been derelict for doing nothing about it. He further charged the Federation did nothing to help several hundred Socialists get their jobs back when the Anaconda Company laid them off after the election of a Socialist mayor in Butte.[4]

This last charge related to the firing of almost five hundred Finnish miners on March 20, 1912. The Finns were part of a general influx of new immigrants into Butte and Anaconda in those years, that included Italians, Slovenians, Croatians and Serbs, as well. The labor press talked alarmingly of cheap labor and referred to the latter as "Dagos", "Bohunks" and "European Chinese". In spite of the difficulties the Southern Europeans had being accepted, one factor eased their passage. They were Catholic and went to the same churches as the long-time resident Irish. But the Finns were another matter, they had become the fourth largest ethnic group in Butte. Clannish because they didn't speak English, they formed their own enclave in a three-or four-block area off east Granite and Broadway, and had little occasion to mix with the larger community. Not only were the Finns not Catholic, a large number of them were Socialists. Most were single, living in large boarding houses that fed three or four hundred boarders a day. Many were experienced miners out of the Lake Superior copper district and excelled in dangerous sinking and timbering operations. They were hard workers, hard drinkers, rough and ready fighters and quick with a knife. The BMU had a closed-shop agreement with the Company, so anyone the Company hired automatically had to join the union. All of the newcomers became union members, however unwelcome. But the union remained firmly in the hands of the old-timers, who knew how to horse-trade with the Company and keep the newcomers in their place.[5]

After Lewis Duncan was elected mayor of Butte on the Socialist ticket in 1911, the Company decided it could not afford to have Socialists controlling the city government, so it got rid of known Socialists on its payroll. It turned out most of the men fired were Finns. It is hard to know whether this influenced the union's attitude in the matter, but it is conceivable that those in control of the union saw a certain symmetry in the firings. The Company got rid of the Socialists and the Union got rid of the Finns. Nevertheless, based on protests within the union, a delegation was named to talk to mine managers about reversing the Company's actions.

The Company admitted that some foremen might have been "overzealous" and offered to reinstate some of the Finns who had lost their jobs. But Thomas Campbell, as head of the delegation, insisted with the union leadership that unless everyone who was discharged were reinstated, the union should call a strike. However, others on the delegation recommended that a strike decision should properly be put to a vote of the entire union membership. The mines were closed down for two days while the voting took place. The strike call was defeated 4,374 against and only 1,126 voting for. The matter was settled, but Campbell and his associates labeled the union leadership and those who called for the vote company stooges and would later criticize Moyer and the WFM for not getting rid of union leaders who were doing the Company's bidding.[6]

There was probably more than a grain of truth in many of the charges that Campbell brought against the BMU at the WFM convention and, had he handled it another way, he might have produced some reform in the BMU. But there was also evidence that Campbell had an agenda of his own, which was not necessarily in the best interests of either the BMU or the WFM. While he denied that he was an IWW agent, he wanted WFM to forcibly change the leadership in Butte, ignore the wishes of a majority of BMU members and instead follow a program which embraced many IWW goals. His charges against Moyer certainly did not further his objective of getting the WFM to take remedial action in Butte but rather appeared designed to cause dissension within the Federation.

Even though Campbell did not have the support of his own union, Moyer and the convention heard him out, then voted to reject his proposals and to expel him from the Federation for having brought charges against WFM leadership which he could not substantiate and which were "designed to poison the minds of the membership against elected officials."[7]

Whether these problems within the BMU were due to the agitation of IWW agents or were just early signs that the union was losing its effectiveness, it signaled the beginning of a rapid decline in membership. In spite of the closed shop, by February 1913, out of a total of 11,000 miners, only 7,000 were members. By May 1914 it was down to a bare 4,000. Not only had membership declined but the number and strength of the dissidents within the union had grown.[8]

Discontent was also on the rise because of increased assessments to support a strike of copper workers in Michigan and rumors about misappropriation of union funds. The impression grew that there was graft at the top, that union leaders were using union dues for their own private purposes. As a result many members quit paying their dues. The situation came to a head on June 12, 1914, the day before the big Miners' Day parade, when BMU set up an inspection of union cards at the mine gates. The night shifts at the Black Rock and Speculator mines refused to show their cards and drove off the union representatives sent to examine them. They then organized a meeting and appointed a committee to inform the mine operators that they would refuse to go to work if they had to have a BMU card.[9]

The next day, while the Miners' Day parade was in progress, a mob attacked the union officers leading the parade, pulling them from their horses. The horses stampeded into the crowd and the entire route of march was thrown into confusion. Someone mounted one of the horses and referring to himself shouted, "How does it look to see an honest man heading a miners' parade?" The parade by this time had been transformed into a milling mass as the mob continued its assaults and the crowds along the route panicked. Then a cry went up, "To the Miners Union Hall, tear it to hell and get the records. Wreck the house of the grafters." And there was a rush toward the Hall. They surged inside wrecking everything in sight and moved upstairs to the union offices where they threw the ballots from the previous election out the window and emptied out the union membership files. Someone was heard to holler, "We're all in good standing now," as the records rained down on the street below.[10]

Mayor Duncan was in Chicago, but the acting Mayor and the Chief of Police soon arrived on the scene. The Chief was hit in the head with a bottle and the acting Mayor, who went into the Hall and out onto a second floor balcony to try to calm the crowd, was pushed off and fell into the street. Luckily, he landed on a pile of carpets and ended up with only a broken arm and a sprained ankle. By this time the rioters were dragging a large safe out of the Hall. A police dray pulled by two strong horses came up the street and the mob took it away from the police, loaded the safe on board and hauled it away.[11]

That night the insurgents held another mass meeting and planned for a referendum of all miners in Butte to decide whether or not a union card should be necessary in order to work. The BMU at the same time was holding a secret session in the Carpenters' Hall where they decided to sue the city for not maintaining order and enforcing the law.[12]

The insurgent referendum was open to anyone who had a union card, no matter how delinquent, a hospital receipt or any other proof that one had recently been engaged in mining. The BMU protested the vote and advised its members not to lend legitimacy to it by participating. The final count was 6,348 against presenting union cards and 243 in favor. This, of course, settled nothing, but was another strong indication that the BMU had lost touch with its membership.[13]

On June 19, Charles Moyer arrived, accompanied by James Lord, president of the mining department of the AFL. He immediately asked for the resignation of BMU officers and appointed William Burns of the Virginia City, Nevada miners union to supervise the affairs of the Butte local pending implementation of a reform program which he would institute. He also called a regular meeting of the union for the evening of June 23. The insurgents, in the meantime, had gone ahead and formed a rival union and elected officers.[14]

Tensions continued to run high in the city and there was a crowd of over a thousand gathered around the damaged Miners Union Hall as the meeting got underway on the evening of June 23. Mike Shovlin of the insurgent union, urged his members to clear the sidewalks and not cause trouble. The president of the new union, rushed to the *Miner* newspaper and struck off five thousand hand bills which said, "Fellow workers in the name of your new union, keep peace and go home. Muckie McDonald, president."[15]

It wasn't to be, however. As Peter Bruneau of the WFM opened the door to go into the hall, a shot rang out, and he fell with a bullet in his shoulder, then more shots. It has never been determined where the first shot came from, within the building or outside. After that, more gunfire was exchanged and Ernest Noy, a bystander, fell dead on the street across from the hall. Two or three others were wounded. During the melee, Moyer and everyone else at the meeting left the hall by a back exit. Later that evening a group of men came back to the hall with dynamite and blasted the building. The dynamiting went on all night with twenty-six separate charges being set off before the building was leveled just before dawn. For miners, supposedly proficient in the use of dynamite, this was hardly a professional job. This gave rise later to speculation that the perpetrators were from out of town.[16]

The following day Moyer left Butte and went to Helena where he met with the governor and described what had transpired. He criticized the mayor and the local police for not preventing the riot. He also blamed the entire sequence of events on IWW agitators. He wrote:

> "The wrecking crew of the IWW is the force at work in Butte...I have positive information that at least 600 IWW agitators have arrived in Butte within the past few weeks... the operations of the IWW are well known. The call goes out for the massing of men in some particular city, and the response is uniformly heavy. They have nothing to lose, and out of the discord and excitement and disorder they get their 'pickins'."[17]

This was the end of the BMU as an effective union force in Butte. While it continued to exist and function with a small following of loyalists, it was challenged repeatedly by the rival union, which was much larger. The Company stopped enforcing the closed-shop agreement that it had with the BMU and by August the new union was forcing miners to join by rounding up and shipping out of town any worker who didn't have his card. Tensions were rising again. Then someone blew up the rustling card office at the Parrot mine. The Company immediately closed a number of mines. Citizens began organizing armed groups to protect property and the sheriff asked the governor to send in the state militia. Martial law was declared and the militia tracked down and arrested the officers of the new union and charged them with dynamiting the rustling office, and the kidnapping and illegal deportation of miners. The Company then declared the end of the closed shop in the mines. Muckie McDonald, president of the rival union, was sentenced to three years in prison, Joe Bradley, vice president, who was believed to be the ring-leader, was given five years.[18]

On the basis of charges made by Moyer and the BMU, grand jury hearings were held concerning the conduct of Mayor Duncan and

Sheriff Driscoll in connection with the events of June 23 and the deportations and "kidnappings" of August 27. Both Duncan and Driscoll were removed from office for dereliction of duty.[19]

Martial law continued in effect until after the elections in the fall and was terminated on November 12. The Socialist administration in Butte had ended. Both the BMU and the rival union were now in ruins. There was open shop in the mines and the miners were at the mercy of the Company. All of this had taken place without any apparent Company involvement.[20]

There was one piece of evidence that suggested otherwise, but it certainly was not sufficient to make a case. When the rival union leaders were being held for trial, there were four men in custody, McDonald, Bradley, Chapman and Shannon. After the McDonald, Bradley trials, Shannon was released. But before the trials began, militia officers not only released Chapman but loaned him money to go to California with his wife, even though he had been conspicuous in the rival union's deportation activities. When this became known there was speculation that Chapman was, in fact, a detective who had been with the Thiele Detective Agency working for the Company. Later during labor troubles in Arizona, Chapman was heard to boast that he had helped put the BMU out of business.[21]

Certainly, the end of the BMU gave the Company a freer hand in Butte. However, it also eliminated a force for stability and discipline among the miners and increased disarray and anarchy among an always very volatile population. With a stable work force of skilled miners, worker productivity had steadily risen. In addition, the BMU leadership and the Company had a long record of mutual accommodation. Even the wholesale firing of the Finnish Socialists had not provoked a strike or loud union protest. Was the elimination of the BMU really in the Company's interest? Probably the view from afar in New York was that it made very little difference either way. The Company would continue to prevail.

26. War Comes to Anaconda

The troubles that the immigrants from the Balkans had sought to escape by coming to America followed them across an ocean, a wide continent and into their new home in the Rocky Mountains. The assassination of Archduke Franz Ferdinand, heir to the throne of Austria-Hungary, on June 28, 1914, at Sarajevo in the distant province of Bosnia, soon had a greater impact on their lives and those of the their Anaconda neighbors than the collapse of the miners' union in nearby Butte.

The Archduke's assassin belonged to a Serbian revolutionary group plotting against the crown. Backed by Germany, Austria determined to put an end to Slavic dissidence and declared war on Serbia on July 28. When Russia mobilized its army to come to Serbia's defense, Germany declared war on Russia. Two days later, Russia's ally France, joined the fray. Germany immediately moved on France through neutral Belgium, and Great Britain, bound in the Triple Alliance with Russia and France, declared war on Germany on August 4. World War I had begun.[1]

From the beginning, the United States was vitally effected by the war, even though the average U.S. citizen had no idea that this country might be drawn into it. Indeed, there was almost universal determination to stay out. This was especially true among the large Irish communities in Butte and Anaconda where anti-British sentiment ran deep. Nevertheless, American public opinion favored the allies, and our financial and commercial interests were closely tied to Great Britain.

Other events on the national scene also influenced circumstances in Butte and Anaconda. The elections of 1912 had brought Woodrow Wilson to the Presidency and a Democratic majority to Congress. This was a national administration of governmental activism. A consensus of Democrats, Progressive Republicans and Socialists pushed through banking, currency and tariff reforms, the Clayton anti-trust law and a graduated tax on income. The excesses of big business also came under greater scrutiny and regulation.[2]

Earlier, Montana had followed the national mood of reform ushered in by Theodore Roosevelt's two terms in the White House by sending reformist Republican Joseph Dixon to the Senate in 1907, where he joined the progressives, was frequently at odds with Senate Republican leadership, and quickly fell out of favor with the Amalgamated Copper Company. Lacking the Company's support, Dixon lost his bid for reelection in 1912. But his successor, liberal Democrat, Thomas J. Walsh, was equally committed to reform. Walsh supported the Wilson program and was a strong critic of Amalgamated's political power in the state. In spite of the Company's hold on local politics, it couldn't buck the national tide of reform.[3]

The reform mood carried over to the 1914 elections when women were given the vote in Montana. This resulted from a whirlwind campaign to educate the public carried out that summer by Mary Long Alderson and a young, energetic, vivacious graduate of the University of Montana, Jeannette Rankin, who two years later became the first woman ever elected to the U.S. Congress.[4]

But even though reform was in the air, news of the war in Europe fed a growing intolerance at home. As the German U boat offensive heated up and American ships and lives were lost, anti-German sentiments intensified and spilled over into acts of violence against anyone perceived to be pro-German or against U.S. support for the Allies. While neither Butte nor Anaconda had a large German population, there was little enthusiasm for Britain either, especially among the Irish and the growing colony of Slovenians and Croats who identified with German objectives. This sentiment, which provided the votes to elect Jeannette Rankin on a "keep the U.S. out of the war" platform, also presented an obvious target for super patriots bent on making trouble. The atmosphere became even more charged as the Socialists and the IWW increased their anti-war activities.

In April 1917, Jeannette Rankin fulfilled her campaign promise and, with forty-nine other members of Congress, voted against declaring war on Germany. For this she was vilified in the Montana press. The Helena *Independent* called her "a dupe of the Kaiser, a member of the Hun army in the United States and a crying school girl". That one vote sealed her fate and cut short what might have been a brilliant political career. When she ran for the U.S.

Senate in 1918, with support from organized labor and the farmer Nonpartisan League, she was soundly defeated.[5]

Once war was declared, America began to mobilize its resources. Many in Europe must have seriously doubted whether this bustling, turbulent nation of immigrants, which sprawled across an entire continent, could organize itself and act in a unified manner. It was not only the question of raising an army, but the gigantic and urgent task of mobilizing vast industrial resources and transforming a highly individualistic economic system into a well-oiled military machine. Fortunately the country was already tied together by telegraph, railroads and telephone. Even the remotest villages could respond promptly to a call from the center of government. Anaconda and Butte typified the integration these advances had brought. Both these nominally remote outposts were connected to the rest of the country by four major railroads and both were soon on the front lines of the war effort. Copper was a strategic resource.

On the national level, Congress passed the Selective Service Act ordering all men between twenty-one and thirty one to register for military service. It also conferred far-reaching war powers on the President to direct national production and distribution, control prices, take over and operate transportation and communications facilities and even expropriate essential industries and mines. These powers were carried out under the Council for National Defense and series of boards in specific areas. The most important of these boards for Anaconda were the War Industries Board under Bernard Baruch, which made the critical decisions on the production, fabricating and sale of copper, and the Food Administration under Herbert Hoover, with authority to fix the price of staples, license food distribution, coordinate purchases, supervise exports, stimulate production and prohibit hoarding and profiteering. The Food Administration would determine what food the Anaconda housewife could put on the table. It introduced "Wheatless Mondays", "meatless Tuesdays" and "Porkless Thursdays" and promoted such novelties as whale meat, shark steak, sugarless candy, potato and rice flour and vegetable lamb chops.[6]

The organization charged with uniting the country idealogically was the Committee on Public Information under George Creel. It had a major impact on public opinion throughout the country, and Butte and Anaconda were no exceptions. A nation divided and in doubt when war was declared was converted slowly into a single voice crying for the defeat of the hated "Hun". Propaganda pamphlets, posters, magazines and newspapers all played the same tune. The *Anaconda Standard* was typical of most daily newspapers in the country in giving no quarter to any critic of the United States or its leaders. An editorial in November, 1917 had this to say: "Every man in America who is against this country and for Germany should be driven out of this land or interned." Aliens must conform to "American ideas, American sentiments, American patriotism, or they must be exceedingly quiet while the war is on...In America at this hour every man must be a patriot. Those giving expression to disloyal sentiment cannot and will not be tolerated." A week before it had said: "Seditious utterances are crimes against the nation. Everyone living in this country must support it." The *Standard* made no mention of first amendment rights to freedom of speech. Perhaps these had been suspended for the duration of the war.[7]

The *Standard* also began publishing in serial form the memoirs of Marcus Daly's son-in-law, Ambassador James Gerard, which gave a strong justification for U.S. involvement in Europe. The Information Committee not only used Gerard's book entitled *My Four Years in Germany*, but enlisted him on a cross-country speaking tour to deliver his message. In Montana, he spoke throughout the state, and the *Standard* gave him banner headlines when he spoke in Butte urging a united effort, asking all to "join hands in the great work". In addition, the Public Information Committee made maximum use of the newest medium of public entertainment, the movies.

By that time there were four movie houses in Anaconda, "The Margaret", still the principal vaudeville house, the "Imperial", the "Blue Bird", and the "Alcazar". Anaconda audiences flocked to silent films featuring Charlie Chaplin, Harry Carey, Mary Pickford, Jack Pickford, Pauline Frederick, Billie Burke, Wallace Reed, Vivian Martin, Charles Ray, Douglas Fairbanks, George M. Cohan, Al Jolson, Fatty Arbuckle, Ann Pennington, Dorothy Dalton and Enid Bennett. They roared with laughter at Mack Sennet's slapstick comedies, and marveled at the spectacular productions of D.W. Griffith and Cecile B. DeMille.[8]

After the U.S. entered the war, Anaconda movie houses played a growing number of new feature films promoting patriotism and support of the war effort, such as Samuel Goldwyn's "For the

Freedom of the World". Audiences were duly horrified by the barbarities of the Huns depicted in newsreels and special propaganda films which idealized U.S. participation in the war and vilified our enemies.[9]

The Selective Service Act was passed in May, 1917. In Butte, on June 5, the day before registration under the Act, the Pearse-Connolly Irish Independence Club, Finnish Socialists and members of the IWW staged an anti-draft parade and demonstration which ended in a minor riot and the arrest of many of the participants. Pearse-Connolly had become the most outspoken and active of the Irish societies in Butte. For its members, Britain was the enemy, and no true Irishman should be forced to fight for her. They joined forces with the Finns, who were sworn enemies of Britain's ally, Russia, and with the IWW, which opposed the war on ideological grounds.[10]

The protest, in which about 70 Finnish women also marched, had little effect on the Selective Service process. The next day 11,603 men were registered for the draft in Silver Bow county and 2,557 in Deer Lodge county. Across the United States total registration exceeded 9.5 million.[11]

It was reported that the figures for the draft quotas would be based on the most recent census, which surprisingly gave Deer Lodge county a population of 24,742, the highest official figure ever used. This, of course, meant that the Anaconda levy would be higher than if the previous census figures of 16,000 or 17,000 had been used. Similar inflated population counts were used throughout the state. Based on these figures, Montana was expected to provide 10,423 men for the initial draft. Anaconda's quota was 246 men. Credit was given for those who had already enlisted, which in Anaconda's case was 160, leaving 86 to be drafted. No one protested the inflated population count. After all there was a war on and Anaconda would do its part.[12]

In the rigorous physical examinations for all those called up, a surprising 77% of Anaconda registrants were declared fit for service. This was significantly above the national average of under 65% and was similar to the high acceptance rates in such states as the Dakotas, Wyoming, Colorado, Nebraska and Texas. The first Anaconda draftees left in four contingents, on September 7, September 24, October 8, and Nov. 7. The first contingent had only twelve members, the second and third had ninety-three each and the fourth had thirty-three. Thus, the town greatly exceeded its initial draft quota, sending a total of 231 men, in addition to those who had already volunteered. Before the war was over a total of 976 Anaconda men were drafted into service and 250 enlisted.[13]

In the spirit of the times, both Butte and Anaconda staged rousing ceremonies for each departing contingent. September saw a flurry of parades in Anaconda beginning with one on Labor Day, September 4, another one September 7 and a third on September 24. The first contingent was honored by the Rotary Club, the Knights of Columbus and the Hibernians which all gave farewell dinners, and by a demonstration and parade from the Montana Hotel to the train station. The first draftees departed for Butte at 6:00 p.m. and proceeded on to Camp Lewis, Washington, to begin military training. The following day, the *Anaconda Standard* ran a front-page photo of the inductees on the steps of the Montana hotel with the banner headline, "Great Wave of Patriotism Sweeps over Anaconda, Cheering Thousands Accompany First Twelve Men To The Train...5,000 at the station."[14]

Subsequent parades started at the Court House after a brief ceremony and then proceeded down Main Street to Park, Cedar, Commercial and back to Main and the train station. A parade for the fourth contingent was typical. They were due to depart for Butte and Camp Lewis at 8:a.m. Monday morning, November 7, so a parade was scheduled for 7:a.m. The reporting time for the day shift at the smelter was changed from 8:a.m. to 9:a.m. and the street car schedule was changed so that men going to work could see the parade and go to the station to see the boys off. Even at that early hour of the morning, crowds lined the streets from the court house to the train station. The parade was headed by an honor guard of military police, Mayor J.A. Hasley and John Marchion, parade marshall. Two members of the departing contingent carried the colors, followed by the Anaconda band and the First Company of Coastal Artillery, which was stationed near town to protect the smelter. Then came the Spanish War Vets, members of the GAR, Elks, Rotary Club, the Cristoforo Colombo society, the Anaconda Bar Association, the Musicians Union band, the carpenters, butchers, barbers and other unions followed by the recruits themselves and the high-school cadets and boy scouts bringing up the rear. The parade had been preceded by dinners given by various civic and fraternal organizations and a big dance given by the Patriotic Ladies Association on Saturday night.

It would be a while still before the Anaconda recruits would be ready for combat. And it would also be sometime before the U.S. would influence the tide of battle in Europe. Though the first American troops of the regular army had arrived in France in June, 1917 only the First Division had seen action by year's end. By that time over 200,000 American soldiers had reached France. The first U.S. casualties occurred in November. Most of the Montana recruits did not arrive until several months later.[15]

On the home front, it took time for the seriousness of the situation to sink in. As the months wore on, Anaconda like the country-at-large increasingly focused on the war and the war effort. Anaconda responded by over-subscribing drives for the purchase of Thrift Stamps and Liberty Bonds and surpassing its quota with new volunteers in each and every draft call.

As the war progressed and local boys names appeared on the casualty lists, Anacondans paid close attention to the latest news from the front lines and the *Anaconda Standard*, proudly touting its *New York Times* Cable Service, dedicated not only its front page but much of the rest of the paper to war news. By April of 1918, it was running the entire daily list of casualties released by the War Department, highlighting any from Montana.

Those who stayed at home did their part producing copper, quietly accepting wartime rationing, and participating in Liberty Bond drives. Many also made generous contributions to the Red Cross, the Knights of Columbus, the YMCA, and the Salvation Army war relief fund drives. A local War Chest fund was organized to which wage-earners made monthly donations. When the war ended the Fund had a surplus of $13,000 which it donated to the returning veterans for a war memorial. In 1926, these funds were used to buy forty-nine musical instruments to fully equip the high school band, which from that time on was called the "Anaconda High School Soldiers Memorial Band." The balance was used to design and build a monument to the soldiers of all wars. This was erected in 1927 at the top of Main Street in front of the Court House.[16]

27. Murder in Butte

When the war broke out in Europe, copper was selling for eleven cents a pound with inventories on the rise and prices dropping. The country had been sliding into recession. Even though munitions makers needed copper, half the mines in Butte were closed soon after the war started. The British were blockading shipments to the continent and yet were short of cash to buy it themselves. As a result, in 1914, production declined (from 230 million pounds of finished copper in 1913 to 197 million), manhours worked in Butte went down 30%, and so did the Anaconda Company dividend.[1]

This situation was eventually reversed when U.S. banks, led by J.P. Morgan, arranged large loans to Britain. This finally created a booming market for copper and other commodities needed to fight the war. By January 1915, copper prices firmed and began to increase. By May there were 11,000 men back in the mines and production by the end of the year had reached 231 million pounds. This coincided with the start of war-time inflation which reversed the long post civil-war deflation that had originally fueled the free silver movement. Demand for copper and other basic commodities was going up and so were prices.[2]

The rise in the price of copper, of course, was good news to Anaconda smeltermen and Butte miners who had contracts tied to price. The copper price index became an item of intense interest to everyone in the two communities and it was posted prominently on the financial page of the *Anaconda Standard*. Smeltermen were by this time being paid $3.50 a day with copper selling at fourteen cents a pound. Wages were calculated on the average price of copper for the preceding month. For every two cent increase in the copper price, wages went up twenty-five cents. By the time the U.S. entered the war in 1917, the price of copper had risen to twenty-eight cents.[3]

So, although the BMU had ceased to exist as an effective bargaining agent for the miners and no longer led interference for the Anaconda unions in their negotiations with the Company, the world situation caused a temporary improvement in the take-home pay of the average worker in both communities. There was not only full employment, but the fixed-term contracts, which the WFM had so vigorously opposed in 1906, insured rising wages, as well. Of course, a good portion of this was eaten up by the rapid increase in the cost of living. In a survey of ninety-two industrial cities in the United States, Butte was one of the most expensive. Also, even with the sliding scale, Butte and Anaconda had lost their reputations, gained in 1900, for having the highest industrial wages in the United States. Butte was far from matching the standard set for unskilled labor in 1914 by Henry Ford of $5.00 a day for an eight-hour shift. High prices, combined with war-time rationing, dampened any feeling among the workers that they had improved their living standard.[4]

By this time, Charles Moyer and the WFM were reevaluating their entire operating philosophy and objectives. Even before the collapse of the BMU, the WFM had changed course. It had joined the AFL in 1911 and was actively pursuing talks with the United Mine Workers of America for the possible consolidation of the two unions. In fact, the chaos in Butte in 1914 caused the coal miners in the UMW to have second thoughts about taking in the WFM. Moyer, for his part, owned up to past mistakes, among which were: staying out of the AFL so long, participating in forming the IWW, opposing fixed-term contracts and union-dues check off, and making unwise declarations in support of socialism.[5]

Nevertheless, the loss of the BMU, the Federation's largest single local, was a blow from which the WFM never fully recovered. Even with the increased war-time activity in copper mining, smelting and fabricating, the federation was only a shadow of its former self. At its convention in June, 1916, at Great Falls, Montana, there were only fifty-two delegates and seven executive officers present, representing thirty-six local unions directly and seven more by proxy. At that meeting, the Arizona locals challenged the Moyer presidency and, receiving support from Butte and Great Falls, almost dislodged him. However, the Anaconda Mill and Smeltermen, breaking with other Montana locals, voted solidly for Moyer, splitting the Montana vote and narrowly assuring his continued leadership. Also at this convention the organization changed its name to the International Union of Mine, Mill and Smelter Workers and amended the preamble to its constitution.[6] The new text eliminated the IWW class warfare slogans and substituted more conventional trade union language:

"We the workers in the metal industry are united for the purpose of increasing wages, shortening hours, and improving working conditions, by removing or providing as far as may be [possible], the dangers incident to the work, eliminating as far as possible, dust, smoke, gases, and poisonous fumes from the mine, mill and smelter; to prevent the imposition of excessive tasks; to aid all organizations of working people in securing a larger measure of justice, and to labor for the enactment of legislation that will protect the life and limb of the workers, conserve their health, improve social conditions and promote the general well-being of the toilers.[7]

These changes had symbolic as well as practical importance. They signified the end of a militant and aggressive labor organization, the Western Federation of Miners, started in Butte in 1893 and dependent upon the large and financially strong Butte Miners Union for its continuing growth and influence. The end of the WFM in 1916 was a recognition of the demise of the BMU two years earlier and a confirmation of Big Bill Haywood's IWW victory over Moyer's WFM. Ironically, the new, smaller and weaker IUMMSW was heavily dependant on a smaller and weaker cousin of the BMU, local 117, Mill and Smeltermen, in Anaconda. This Anaconda union, with the Stationary Engineers in Butte and the Smeltermen in Great Falls, became the core of the new international. John Rankin of Anaconda was elected Vice President and played a central role in its operation through the strikes in 1917, 1918 and 1919. Later in the anti-labor atmosphere of the 1920's he was instrumental in keeping the organization alive to take advantage of the New Deal reforms of the 1930's.

But the changes voted at Great Falls had little relevance in Butte, where the demise of the BMU had introduced a large element of instability into an always volatile environment. The new IUM-MSW, had practically no influence on the remaining local unions in Butte. To make matters worse, even without a dominant union, (most miners not belonging to any union) the old inter-union rivalries continued. More serious still, worker discontent with management was on the rise although the mine owners seemed unaware of it or at least unconcerned. Charlie Clark, W.A. Clark's son, probably spoke for all of the mine owners, including the Anaconda Company, when he wrote in 1916, "I am reliably informed that Butte has never been so well satisfied, so far as labor conditions are concerned, as in the last two years; that is since the elimination of the union."[8]

Typically, Charlie Clark didn't know what he was talking about because Butte, which had not seen a strike for twenty-six years when Daly was alive, was about to be plunged into a prolonged period of labor unrest and violence, which coincided with the country's mobilization for war and therefore drew national attention and prompted armed military intervention.

On the night of June 8, 1917, a fire in the Speculator mine killed 164 miners. This was the worst mine disaster in Butte's history and brought out all the pent-up anger, frustration and discontent against the Company that had been simmering since 1914. Even though the mine did not belong to the Company, but to the North Butte Mining Company, there was a Company connection, John D. Ryan was a major stockholder.

A strike was called and a joint strike committee, representing the various trades, was formed. The strikers demanded the dismissal of the State mine inspector, observance of mining laws, abolition of rustling cards and an increase in wages tied to the cost of living. The strike then spread to the Montana Power Company, called by the local electrical workers. On June 10, one hundred and twenty-three men of B Company Montana National Guard arrived, bringing the number of troops in Butte to two hundred.[9]

There was no single voice to speak for the disparate groups on strike. A hurried attempt was made to organize a new union, the Metal Mine Workers Union. But it was immediately suspect because its leaders had connections with the IWW, the same Tom Campbell, Joe Shannon and John Shovlin, who had challenged the old BMU in 1913. Moyers, Rankin and their IUMMSW colleagues worked hard to counter the new unions actions, though they had little influence. Company proposals accepted in some quarters were turned down in others. Proposals for settlement put forward by one group were rejected by another. There wasn't even agreement among those striking as to what the issues were. Not all of those on strike, for example, considered the rustling card a major problem.

During June and into July there was continuing confusion and endless delays in bringing labor and management together and the discontent spread to Anaconda and Great Falls. On July 25, the Company announced a new sliding scale of wages raising miners pay to $5.25 a day, but they refused to give ground on the rustling card. While some groups were prepared to accept these terms, the majority refused and the strike continued. However, most of the craft unions and even the electricians, whose leadership was under IWW influence, appeared to be softening. They were aware that with the country at war the government could not tolerate prolonged work stoppages and the consequent loss of production. This, combined with the jingoism of superpatriots on the one side and anti-war demonstrators on the other, contributed to an increasingly volatile atmosphere.[10]

Into this highly charged situation came one Frank Little, an IWW agitator and an anti-war activist. The IWW was strongly opposed to U.S. involvement in the war, speaking out publicly against it and engaging in guerrilla tactics at the union level to create chaos and shut down production. They were already busy in Butte in anti-draft activities and the Speculator tragedy provided an even more inflammatory platform from which to oppose the war. Little had arrived from Arizona where the IWW had taken advantage of worker discontent to promote strikes in Jerome, Miami, Globe and Bisbee. The situation in Butte was ripe for IWW tactics to sow the seeds of even greater confusion.[11]

On July 19, the day after Little arrived, he made an anti-war speech which many considered seditious. Among other things, he called American soldiers "scabs in uniform". The *Anaconda Standard* reported that he said in a speech, referring to troops in Butte, that it was time to get tough, "the capitalists are using their rifles, working men should use theirs".[12]

At three o'clock in the morning of August 1, six masked men drove up to Little's rooming house and hustled him outside, not even allowing him to dress. They tied him with a rope and dragged him through the streets behind their car. In the morning his body was found hanging from the trestle of the Milwaukee railroad where it crosses the Butte-Anaconda highway. Printed in red letters on a card pinned to his underwear was the message, "Others Take Notice. First and Last Warning. 3-7-77." The numbers denoted the measurements of a grave (three feet by seven feet by 77 inches) and were the old signature of the vigilantes of Bannack and Virginia City days.[13]

IWW activism had led to a tragedy of major proportions. It was assumed in most quarters that Little had been killed for his seditious speeches. The IWW answered, however, that if his speeches had been seditious, he would have been prosecuted by the U.S. government. They described the murder as "work of the gunmen of the Copper Trust." The *Miners' Strike Bulletin* even ran the names of five individuals who allegedly participated in the killing. However, no evidence was forthcoming and when questioned by the county attorney, the IWW lawyer would not state that those named had been involved. Other rumors circulated that Little was employed by a detective agency that had infiltrated the IWW. This charge Charles Moyer and his ailing IUMMSW were prepared to believe, since they were convinced that the IWW was a tool of those who wanted to wreck the labor movement.[14]

Whatever the truth, the murderers were never discovered and Little became a martyr memorialized in the IWW song book. The lynching was widely criticized. Even the *Anaconda Standard,* which had given front page space to stories from all over the United States about actions taken against IWW agitation, and for some time had dedicated at least one editorial a week to castigating the IWW and criticizing the U.S. Attorney, Burton K. Wheeler for not prosecuting IWW activists, condemned Little's murder in the strongest terms. In a banner headline the morning after the murder it asserted: "BUTTE'S NAME TARNISHED BY STAIN OF LYNCH LAW".[15]

Little's funeral was one of the largest Butte had ever seen. The *Standard* reported on it in detail describing the procession, a four-mile journey from the Federal building on north Main Street to the Mountain View Cemetery. There were 3,500 marchers in the procession and over 10,000 spectators crowding the streets. The largest contingent represented the Metal Mine Workers Union, followed by the Electrical Workers, the Hod Carriers, Cement Workers, The Pearce Conally Club, 120 strong and over 200 women, most of them Finlanders. The coffin was carried through the streets on the shoulders of six strong miners and was followed by a twenty-man guard of mourners. The *Standard* commented that the mourners guard was made of "mostly IWW from out of town, Finns, Austrians, Bulgarians, and Turks". All these nationality groups were on Germany's side in the war. The procession was peaceful and without incident watched over by the entire contingent of the Butte police force. Several days after the funeral more federal troops moved into Butte. The city was once again under martial law.[16]

As the strike impasse continued, the situation deteriorated. Federal troops protected the mines and patrolled the streets. Striking miners attacked miners who tried to return to work. Vigilante organizations attacked striking miners and their supporters and questioned their patriotism. W.A Clark declared that he would rather flood his mines than recognize the union. The miners called a mass meeting at Columbia Gardens, to petition the government to take over the mines "so that miners may give prompt and practical evidence of their patriotism".[17]

Congresswoman Jeannette Rankin was invited to Butte to address the striking miners. She was still a very popular figure in Butte and Anaconda and much was made of the fact that she was the only woman in the U.S. Congress. She arrived in the evening and was met at the train station by a large crowd. Butte officials were nervous that her presence might set off a demonstration so had prohibited a parade and insisted that she address the miners at the ball park at Columbia Gardens the following day. The crowd waiting at the station chanted, "Parade, parade, to hell with the police." But Rankin quieted them down and told them to come to the ball park the next day.[18]

In her speech, she didn't mince words. Criticizing both labor and the Company, she said, "I have no patience with the spirit which seeks to destroy property to satisfy personal grievance..." and she assailed "character assassins who would poison the minds of their fellowmen with falsehoods...I have no patience with the alleged utterances of Frank Little...but his death was foul murder." She went on:

> "Let it be known that it is unpatriotic for capital to refuse the demands of labor and it is unpatriotic for labor to refuse the just demands of capital...In short, if the Anaconda Copper Mining Company continues to hold up the production of copper from the country in time of war, it does it simply because it obstinately refuses to abandon the rustling card."[19]

She added that if the owners could not operate the mines the government could.[20] Of course, it was an oversimplification to blame the entire situation on a single issue about which there wasn't even agreement among the groups on strike. But there were many beside Rankin who thought that controversy over the rustling card was at the bottom of the miners' discontent.

The pastors of the First Presbyterian Church in Anaconda and the People's Community Church in Butte, concerned by the lack of progress in settling the strike, issued a statement reminding Company representatives that it was the tragedy at the Speculator mine which had precipitated the strike, that all of the miner's grievances against the Company were there before the IWW had come on the scene, and that the rustling card was the original grievance which led to the "disruption of the old miners union". The Metal Mine Workers Union claimed that the rustling card was, among other things, a device for forcing miners into hot and unsafe mines and that the Company had denied rustling cards to safety-conscious veteran miners who protested unsafe conditions in the mines. They claimed over 3,500 men had been refused cards, most of whom were not drifters, but substantial, property-owning members of the community. The Company was thus blacklisting the very men who had helped build the city.[21]

By mid-August the troubles had spread to Anaconda where there was growing agitation for the smeltermen to join the strike. In July, Tom Campbell had gone to Anaconda to assess the situation, but he was met at the train depot by the Sheriff and the Chief of Police and told to return to Butte unless he wanted to spend the night in jail. This was not an unusual procedure. Ever since the Speculator tragedy, suspected IWW organizers were met at the depot and escorted to the Sheriff's office where they were searched for incriminating material. If any was found they were invited to leave town immediately. But Campbell was different. He had many friends in Local 117. When they heard about it, the local issued a resolution condemning city and county officials for their actions and inviting Campbell to a meeting on August 2. So, in the midst of the flurry over Little's murder, Campbell returned to Anaconda. He was greeted at the depot by a crowd of 2,000 which grew to 3,000 as they walked from the depot to the union hall. While Campbell urged the union to join the strike, his words were not inflammatory as Little's had been. He cautioned the men to "stay within the law" and "if things get too bad, take a few days off and go fishing."[22]

The Company had already announced that it would not recognize Campbell's union as a bargaining agent and John Rankin and William Davidson of IUMMSW were in town lobbying against a strike vote, reminding the men that there was a war on and that the Company offer was fair. However, they were concerned enough about the possible outcome of a strike vote to remind union members through a notice in the *Standard* that the constitution of the international required that two-thirds of the total union membership vote in favor of a strike to make it effective.[23]

Meanwhile, the Company had already taken the precautionary step of constructing a fence around the smelter. Troops had been guarding the perimeter since May. Local 117 took a strike vote on August 14. The union split down the middle with 1,008 in favor of the strike and 1,072 against. Since the strike vote didn't carry, the smeltermen stayed on the job, but the situation remained tense.[24]

Tom Campbell and his fellow militants were not to be deterred, however. For them the vote merely set the stage for their next move. They immediately set about organizing a chapter of the Metal Mine Workers union in Anaconda, renting space in the Austrian Hall and began raiding Local 117 and the trade unions of their members. Campbell was serious about encouraging the smeltermen to "go fishing". To make it easy for them to do it, he set up a picket line at the smelter and threatened physical violence to anyone who crossed it. Over the next several days the number of men reporting for work fell off considerably and on Aug 24, only 110 men out of a work force of 3,500 showed up. So the Company closed the smelter, its mines in Butte and the refinery in Great Falls. The *Standard* reported the next day with a banner headline, "Black Friday, Smelter Forced to Close Down. Radicals Win Out. 15,000 Men Idled."[25]

Campbell and his Mine Metal Workers had finally stumbled onto the tactic which could accomplish their goal, to close all the mines and shut the Company down. They had tried since even before the Speculator fire to get all of the Butte miners to go out on strike thus halting the production of copper at the source but they had failed. They were never able to win over more than one-third of the miners. Now, however, by establishing an effective picket line in Anaconda, the Company could not process the ore being mined and was forced to close not only the smelter but the Butte mines and the Great Falls refinery as well. Shades of 1903. Campbell, Shovlin and cronies must have been dancing with glee. But the fight wasn't over.

Moyer and Rankin in Denver were furious when they learned of Campbell's maneuver. They were witnessing a replay of the destruction of the BMU in 1914. Immediately they sent a communique to all of their affiliates in the United States which read in part:

[When the strike began in Butte after the Speculator fire] "approximately 12,000 miners had been employed." [Until the shutdown], "6,000 continued to work", as well as those in Anaconda and Great Falls. All of the men have two year contracts with the companies. "Approximately 3,000 miners have left Butte because of the strike, leaving only about 2,500 men in Campbell's IWW union. In other words, 15,000 men in Montana affiliated with AFL have signed and are working under satisfactory contracts, while less than 3,000 refuse to accept contracts and are asking for help from bonafide labor movement outside Butte to contribute to their support and to help to feed and fatten a large number of camp followers of the IWW, who have hastened there to hasten the destruction of the bonafide organization." No true unionist should be fooled and contribute to their cause.[26]

This statement, while a good summary of the true situation in Butte and Anaconda, also illustrates the weakness of the IUMMSW. All it could do was plead that its members and other AFL affiliates not support Campbell's actions in Anaconda.

The Anaconda smeltermen were not prepared for a long work stoppage. It was summer and a couple weeks of fishing was fine, but what about the long haul? At the same time the first contingent of draftees were being called up to serve their country, to make, perhaps, the ultimate sacrifice. Outright opposition to the war was minimal among the workers. How could they justify refusing to make some sacrifices of their own? The situation was not propitious for a long strike and the Metal Mine Workers' rump union in Anaconda had a hard time holding the line. Not only were the Company and the IUMMSW working against them, but Federal Mediator W.H. Rodgers let it be known that a strike would violate both the charter and constitution of the IUMMSW and as such could not be sanctioned by the government.[27]

Traditionally, Dame Anaconda always held her passions in check better than her older brother, Butte. Perhaps this was because she was usually slightly removed from the center of controversy. The present situation was no different. As incongruous as it seems, in the midst of all the confusion the Anaconda unions were preparing for a huge Labor Day celebration to take place in Anaconda on September 3, at which Butte unions were to be honored. This may also give some inkling how much priority a strike had in the eyes of the rank and file.

On the appointed day visitors from Butte arrived by train, car and bus to take part in the festivities. These included a parade through the center of town ending at the city common where a baseball game between the best of the Butte and Anaconda unions was played and then on to the Washoe park for a picnic. The celebration was declared a great success by all who attended and the Anaconda unions were especially jubilant that their team had beaten Butte, 15 to 8.[28]

But the propaganda war was heating up. The *Standard* published a long anonymous letter signed "America First," which attacked Big Bill Haywood and the IWW and linked them directly to the shutdown in Anaconda. It reminded readers of the murder of Governor Stuenenberg, the Haywood murder trial and the testimony of Harry Orchard recounting his and IWW inspired crimes. On September 10, the effigy of a dead miner covered in blood was found hanging from the cross-arm of a telephone pole near the Silver Lake camp ground west of Anaconda. A printed message was attached to it which read: "The remains of the SOB IWW lies somewhere in the lake, where will follow Campbell, Dunn, Shannon, Sullivan, Shovelin, Tomich and Robinson and two from Philipsburg, Wm. Yarnell and Arch Owens. The above is a dirty bunch of traitors and their names are good and plain. First and Last Notice."[29]

By this time there had been informal talks among the Company and the various union heads and a meeting was scheduled for September 11, with Local 117, what was left of the BMU, and all the craft unions affiliated with the AFL at the Margaret Theater. The company proposed a fifty-cents a day increase in wages and offered some concessions on working conditions once again, but gave no ground on the rustling card. However, since this had never been an issue in Anaconda, the offer was accepted by unanimous acclamation. More federal troops were assigned to Anaconda for "patrol duty" and to "quell any disturbances connected with a return to work Sept. 17". On September 18, the smelter fires were stoked anew and the mines in Butte began operations. Tom Campbell's group held out until December 28, when their strike was called off. The rump union set up in Anaconda did not prosper and soon withered away. Martial law, however, remained in effect, with Captain Omar Bradley stationed in Butte at the head of the peacekeeping force.[30]

Even after Campbell's union called off its strike, the IWW kept up a drum beat of agitation, trying to increase membership in the union. In July 1918, it presented to John D. Ryan, who was visiting Butte, the same demands made during the strike the year before. Cornelius Kelley replied in a public statement that the Company would never deal with the IWW. In September several thousand miners walked off the job demanding abolition of the rustling card system and a basic wage of six dollars for an eight-hour day. No other union endorsed the strike. Federal troops raided IWW headquarters and the offices of the *Butte Bulletin* and took twenty-four men into custody. By the end of the month the strike had fizzled.

The *Anaconda Standard*, which had been conducting its own editorial campaign against the IWW, made sure to publish any news that put the organization in a bad light. Thus on November 16, 1918, it carried a long article about a meeting of labor leaders at the Pan American Labor Conference in Laredo, Texas. The Mexican delegates to the meeting had introduced a resolution calling on the AFL to work for the release of IWW leaders jailed under the Espionage Act. The reaction from Samuel Gompers was energetic and direct. The *Standard* quoted part of his speech, "The IWW in the United States is just exactly what the Bolshevicki is in Russia and we have seen what they have done to the working people in Russia, where people have no security, no peace, no land and no bread." Even more to the point was the speech given by Charles Moyer, President of Mine, Mill and Smelter Workers International and reflected his long feud with Big Bill Haywood:

"My name is Moyer. Remember that name and go back to Mexico and tell them what I say. Twelve years ago Moyer, Haywood and Pettibone were on trial for their lives and although we were not affiliated with the American Federation of Labor, that organization contributed the funds which enabled us to prove our innocence and saved our lives.

"Pettibone died, and on his death bed he gave thanks to the American Federation of Labor for the help it had given him. Moyer lived and served and tried to be true to the organization to which he owes his life. He is now paying part of the debt he owes them.

"Haywood was false. He started at once an organization to undermine his benefactors and injure his country. I warn you to go back to Mexico and tell the miners that you have learned from Moyer to look and be careful, and when you find a man representing the IWW to know he is there not to organize you to better your conditions, but he is doing there what he is trying to do in the United States—to organize you to destroy you."[31]

The Mexican resolution was defeated, and while IWW influence continued in Butte for the next couple of years, Company strong-arm tactics and aggressive federal actions broke their back. Federal troops and special deputies broke up a strike after eleven days in February, 1919 called by the Metal Mine Workers to protest a reduction of $1.00 a day in wages in Butte, Anaconda and Great Falls following the postwar collapse of the copper market. Then in August 1920, during another strike gunmen fired into a group of pickets on Anaconda road in Butte killing two men and wounding 19. While it was widely assumed that this was a Company inspired act, no one was ever prosecuted.[32]

What was happening to the IWW in Butte merely echoed events in the rest of the country. In 1918, in Canton, Ohio, Eugene Debs was sentenced to ten years in jail because his speeches were held to undermine the country's faith in the draft. And in Chicago, Big Bill Haywood and thirty-eight other members of the IWW were charged with "10,000 specific crimes" of speaking and writing against the war and sentenced to jail terms from four to twenty years. In the hostile anti-labor environment of the 1920s the IWW faded away. Moyer and his IUMMSW came close to suffering a similar fate.[33]

28. *"Please Do Not Congregate."*

The Anaconda Company worked in close collaboration with the government throughout the war. Ryan became the unofficial spokesman for the U.S. copper industry and dealt directly with Bernard Baruch, the chairman of the War Industries Board, in establishing production goals and prices for copper. Initial prices for sales to the government were set at 23.5 cents per pound in September 1917, even though wages continued to be paid at the market rate of 27 cents. The price was subsequently raised to 26 cents in July 1918. To meet war-time needs the Company also began smelting and refining zinc and producing manganese from the Butte ores.[1]

Ryan left the Anaconda Company temporarily when he was appointed Assistant Secretary of War in May 1918 and put in charge of aircraft production. The U.S. had entered the war the year before with only a handful of obsolete planes and American aviators had to depend on European models until domestic production increased. A massive building program was needed and Ryan began it. He had put in place arrangements to produce 23,000 planes a year just as the war ended in November.[2]

As the war-time demand for copper grew, the capacity of the smelter increased and the number of its activities expanded. The three-hundred foot smoke stack that had been constructed in 1903 could no longer handle all the smoke and flu gasses, so it was decided that a new and larger stack must be built. The concrete base was completed on May 16, 1918 and on May 23, there was a ceremony at which Frederick Laist laid the first brick for the new stack. Acid-proof bricks specially made for the stack were manufactured locally at the Company's brick plant. The enormous octagonal concrete base alone required 21,000 sacks of cement, fifty railroad cars of sand and 250 cars of crushed rock.[3]

Work proceeded rapidly, with three shifts a day of a dozen bricklayers a shift, working on special scaffolding erected inside the growing structure. By August 20, the stack had reached a height of two hundred feet and by the middle of September it was taller than the old stack. The last brick was laid on November 30, 1918, just nineteen days after the war ended. Inside the flues at the base of the stack, hundreds of miles of wire were strung as part of a system for capturing particles of toxic dust on electrostatic precipitators called Cottrell treaters. The entire job had taken only 142 working days.[4]

Now the city had a new symbol, standing 585 feet, 1½ inches tall, it was the tallest smoke stack in the world, taller than the Washington Monument. Situated on top of the highest smelter hill, it towered over the town and the surrounding valley. Like an ancient cathedral it stood out like a beacon. Approaching Anaconda from any direction, this was the first sight that caught the eye. Could there possibly be a stronger affirmation of the town's continuing worth and of its bright future? As long as there was smoke coming out of that stack, there would be life in the town.

Once again the Company broke out the champagne and threw a big party. Elevators inside the stack took people to the top where a gigantic platform had been constructed. An orchestra was on hand and couples danced the afternoon away on a dance floor in the sky. Matt Kelly says that while a party of New Yorkers was at the top, George Schutty, Assistant Fire Chief at the smelter, climbed the outside iron ladder the entire 585 feet and came over the top onto the platform. Several of the ladies seeing him come over the top fainted from the shock of it. After shaking hands all around and giving a few autographs, he left the same way he had arrived.[5]

As work on the stack proceeded the *Anaconda Standard, using its* New York Times cable service, bombarded its readers daily with war news. The front page, day after day, carried dispatches from the front lines in France and Belgium and commentary on the war from all the international capitals, crowding out all but the most catastrophic local news. The inside pages published the daily War Department list of casualties, which on a given day might occupy two full pages. Separate boxes listed current Montana dead, wounded and missing in action. In addition, the Anaconda and Butte sections of the paper carried Honor Rolls of names of all the dead and wounded from Anaconda and Butte since the U.S. entered the war. The progress of the war was the focus of interest and the central topic of most conversations. The tide had turned and the Allies were pushing the Germans back on all fronts.

14. The Washoe Smelter and the "big stack" constructed in 1918. Smoke from the stack signalled work and wages for the town. No smoke meant hard times. Note fence around the perimeter (dark line) built at time of labor unrest during WWI.

Under the circumstances, few people paid attention to a brief item out of Boston, Massachusetts, in the September 18, *Anaconda Standard* which ultimately would add to a roll of deaths throughout the U.S., including the state of Montana, even, more extensive than those inflicted on the fields of battle. It read: "Army Camps Invaded by Spanish Influenza" and described an epidemic of flu in Camp Devons, Massachusetts, Camp Upton, New York and Camp Lee, Virginia. It reported that there had already been seventy deaths in New England from the disease. Subsequent stories in the following weeks confirmed that the nation was in the midst of a pandemic of gigantic proportions taking major tolls in Army camps but gradually spreading to civilian population centers. Day after day the reports grew more alarming: September 26, "Between 50,000 to 70,000 cases in New England; September 27, "Massachusetts in Disease Grip"; September 28, "Seven Thousand Cases of Influenza in Army"; "Epidemic of Influenza In Training Camps Causes Cancellation of Draft Calls in Butte and Anaconda".[6]

There had been mild outbreaks of influenza in both Anaconda and Butte the preceding winter and spring, and even a few deaths, but up to the first week in October, the effects of this new and apparently more virulent disease had not reached southwestern Montana. The first reports came from the eastern part of the state along the route of the Great Northern Railroad, Libby, Choteau, Scobey, then Billings, and Miles City, and closer Twin Bridges, Whitehall. On October 5, public health authorities in Butte began talking about control measures that might be put into effect should the disease strike, but stating that so far the city had escaped its ravages.[7]

By October 8, new cases were being reported from widely diverse locations, such as Philipsburg, just west of Anaconda, and Poplar. The State Board of Health in Helena issued regulations for the control of Spanish Influenza, giving local health officials authority to close schools, churches, movie houses and other places of public gathering and prohibiting political rallies and similar public events. There were twenty-five cases reported among soldiers at Fort Missoula out of a total contingent of just 206. Sheridan county, in the northeast corner of the state, was the hardest hit with one thousand cases reported, two hundred and six in Scobey alone. The next day the State Board of Health issued a call for doctors and nurses to fight the disease.[8]

With influenza finally classed as a reportable disease on October 8, the first cases were reported in Butte the next day, and the following day authorities ordered the closing of schools, places of entertainment, and churches and the prohibition of all public gatherings. Streets in the business district were to be flushed by water trucks twice a day. Street cars were to keep all windows open at all times to allow maximum circulation of air. The season's first football game for Saturday October 12, between Butte's Catholic Central High and Helena's Mount St. Charles High was canceled.[9]

This prohibition of public gatherings was a hard blow for the Butte Italian community which had been organizing a large celebration for October 12, Columbus Day, with a big parade through the business district and a gala with athletic events and dance at Columbia gardens. They had been delighted when President Wilson declared October 12 a double holiday celebrating not only the discovery of America but Liberty Bond day, the wind-up day for the fourth Liberty Bond drive which had begun October 1. Their parade was to have been much larger and more spectacular than usual. Now everything had to be cancelled.

The Italians weren't the only ones affected. On October 10, with only 15 cases reported in Butte, merry makers discounting the ferocity of the disease and determined to celebrate to the last possible moment, packed into theaters, movie houses, saloons, cabarets and dance halls, all of which were scheduled to close at midnight for an indefinite period. The first death from influenza was reported the next morning.[10]

Protests were heard from various quarters about the unfairness of the quarantine. Representatives of the Women's Christian Temperance Union wrote letters to the editor demanding the closing of all saloons, inasmuch as all churches had been ordered closed. The Saloon keepers for their part tried to allay criticism by offering their full cooperation to health authorities and volunteered to discourage crowding in barrooms.

The *Anaconda Standard* defended the measures taken by local authorities and editorialized that Butte was only carrying out the health guidelines issued in Helena and following the prudent example of cities all over the country. By October 11, the state health board announced over 1000 influenza cases in Montana and twenty deaths.[11]

Developments in Anaconda paralleled those in Butte, even though only four cases of flu had been reported. The Liberty Loan parade for Columbus Day was cancelled, schools, churches, movie houses, theaters and public halls were closed, saloons and pool halls were urged to let no more than six people congregate at any one time, and it was announced that a long dormant city ordinance prohibiting spitting on the street would be vigorously enforced. An official notice was published declaring Spanish Influenza contagious, communicable and dangerous to public health and gave detailed instructions for those infected by it, including rules for sterilization of contaminated areas.[12]

Dr. J. M. Sligh, the city health officer, declared that the extreme measures enacted in Butte, such as prohibiting all public gatherings, might not be necessary for Anaconda, so the season opener football game between Butte High School and Anaconda High on Saturday October 12 at the Washoe Park would be played as scheduled. This decision turned out to be especially gratifying for

Anaconda boosters who saw their team win 3 to 0, when George Hartsell drop-kicked a field goal in the game's waning moments.[13]

During the following days, the number of people infected grew rapidly in both cities. On October 12, Butte reported seventy-one cases, one day later 164, and on October 15, 341. Anaconda reported thirty-four cases on October 14 and sixty-seven by October 17. Then came the inevitable death notices. In Anaconda, Mrs. Alice Beckman died of the disease on October 15, and on the following day Mrs Elsie Schroeder Thompson, born above her parents' bakery on east Commercial Street in 1887, died leaving her husband W.A.J. Thompson and three young sons. Anaconda natives in the service were succumbing too, Dennis McCarthy, born in Anaconda in 1892, died at Jefferson Barracks, Missouri on October 17. As the number of deaths mounted, public funerals were prohibited, no more than six conveyances could be used in a procession to the cemetery.[14]

On October 19, Anaconda closed all pool rooms, card rooms and clubs and made it illegal for more than six people to congregate in a saloon at any one time. Saloons and grocery stores had large signs that said, "Please Do Not Congregate. Transact Your Business and Keep Moving." Workers at Anaconda Red Cross began wearing gauze masks and issued instructions to the general public on how to make them. The *Anaconda Standard* ran a story that the Red Cross was distributing gauze face masks in Washington, D.C. and that government workers were wearing them on the street. The Anaconda health board closed three saloons for not observing regulations and it was decided that liquor could not be dispensed by the glass but sold only by the bottle. All persons dealing with the public were ordered to wear gauze face masks. In order to care for the growing number of victims of the disease, the County government and the Anaconda Company hastily built a new isolation hospital behind the court house and inaugurated it in November.[15]

The disease was raging across the United States. Virginia reported 200,000 cases, Connecticut 110,000 cases, New Orleans 10,000 cases, Minnesota 9,000 cases, and 4,000 new cases in Philadelphia. Queens, New York, where over 2000 victims lay awaiting burial was advertising for grave diggers to inter bodies. The National Association of Motion Pictures decided to suspend all distribution of films because of the flu.[16]

In Montana the number of cases climbed rapidly. Butte alone reported 2,200 cases by October 22, and 1,556 additional cases the following week. The number of new cases grew so fast that Butte authorities converted the Washington Junior High School into an emergency influenza hospital. Deer Lodge converted a hotel into a hospital for the same purposes. Deaths, too, were on the rise. Sheridan county reported twenty dead by October 15. By the end of October Butte reported 305 dead. The columns of death notices for Butte in the *Anaconda Standard* got longer as the month progressed, running between 20 and 30 every day. The Anaconda list was mercifully much shorter with only 2 or 3 a day. While the number of Anaconda cases being treated reached 137 on October 25, Dr. Sligh observed that the town was extremely fortunate that many were recovering from the disease.[17]

This was a plague of major proportions cutting a wide swath through the country's population. People were dying by the millions. In Butte, double and even triple deaths in one family were becoming common. Patrick Sullivan of Butte died on October 26. His wife and three children were all in serious condition. The next day his fourteen-month old son died. The same day his first cousin Nellie Sullivan collapsed on the street and died within two hours. The following day her brother, Patrick Casey, shift boss at the Anaconda mine, died of the disease. The week before there had been a triple funeral for John O'Meara, his wife and daughter, all victims of influenza. In Bozeman, L.E. Brown, father of five children who had died of the flu, also succumbed. His wife died two weeks later. In the little mining town of Neihart, Montana, population 200, where three quarters of the town were quarantined with the flu, there were thirteen deaths in twelve days.[18]

Additional control measures were put into effect, hotels and lodging houses were ordered to undergo regular fumigation, homes with influenza victims were fumigated and posted with placards, federal troops, brought in because of the labor troubles, assisted local police in discouraging crowds on the streets. The usual raucous Halloween celebrations were curtailed and the crowds of little masked "mummers" going from door to door were forbidden.[19]

Names of the dead, either from the flu or casualties of the war dominated the columns of The *Standard*. Neither respected rank or prominence. November 2, All Souls Day, the day of the dead in many cultures, Anacondans were mourning the death of Dr. Ernest Beal, dentist and football coach, killed in action and that of Deer Lodge County Deputy Sheriff Victor Carlson who had died of the flu.[20]

In terms of new cases reported, the disease reached its peak in Butte the last week in October. The week ending November 1, the number of new cases was down to 749 and it declined every week after that. This decline prompted precipitous action on the part of local authorities, under considerable pressure from business interests, to ease the restrictions of the quarantine. On November 8, the Silver Bow County Health Officer J. B. Freund announced that church services could once again be held, the ban on saloons, theaters, movie houses, and dance halls was lifted, and federal troops would no longer be used for crowd control. It was decided that the Washington Junior High's use as a hospital would be phased out so that it would be ready when schools were opened.[21]

But the county health board was not unanimous on these measures. Some argued that the disease continued to claim new victims every day and the number of deaths were increasing rather than decreasing. It was decided that schools were to remain closed, funeral processions would continue to be limited and the decision on Washington Junior High was reversed. Then on November 28, in view of the continued persistence of the disease, the board reversed itself and reimposed restrictions on public gatherings, dance halls, saloons, cabarets, pool halls, card rooms, etc. On November 30, all places of business except grocery stores and restaurants were ordered closed and church services were once again prohibited. Total deaths in Butte for the month had grown to 326.[22]

The course of the disease in Anaconda was similar to that in Butte with the number of new cases peaking the last week in October. But, miraculously, Anaconda was spared much of the havoc the disease wreaked in Butte and other Montana cities. By November 11, just 260 cases had been reported with only eleven deaths. By November 28, the total number of cases rose to 372 and the number of deaths to twenty-four. The Anaconda Rotary Club made a formal declaration noting that "the city of Anaconda has suffered less in a marked degree than many sister cities in the West" and attributed this to the vigorous and ceaseless efforts of the Deer Lodge County Board of Health and especially to Dr. J. M. Sligh, County Health Officer.[23]

While the strong quarantine measures introduced by the Board of Health undoubtedly played an important role in containing the disease, they also worked hardships on the community. With so many community activities curtailed, there was a certain restiveness among townspeople, a feeling that the measures were too extreme for the circumstances. Certainly, the situation was much less critical in Anaconda. As in Butte, Anaconda business men wanted to ease the restrictions as soon as possible. But the Board didn't immediately follow Butte's lead in lifting the ban on church services, public meetings, theaters, movie houses, saloons and dance halls the first week in November.

Perhaps one reason for this was an incident which occurred on November 1. To the surprise and to the amusement of many in the city, the Reverend J.B. Pirnat was arrested for holding church services in violation of county health regulations. What a juicy scandal this was! Father Pirnat, native of Slovenia and rector of St Peters church since 1897, a prominent member of the community, placed under arrest and ordered to appear in court the next day.

November 1, All Saints Day, is a holy day of obligation on the Catholic calendar. In Slovenia it is celebrated as the "First Christmas" mass; a very important church day, indeed. And by coincidence it was also Father Pirnat's birthday. A few days earlier, the good Father had gone to Dr. Sligh and explained to him the particular importance of All Saints Day and asked him to make an exception for just this one time. He offered to hold the mass outside so as not to violate health regulations. But Dr. Sligh insisted that no exceptions could be made and that even outside, no more than ten people could congregate. This was news to Fr. Pirnat and the rest of the town for that matter because no regulations on congregating in the streets had ever been issued. One can imagine the conversation going something like this:

Fr. Pirnat: Are you forbidding me to hold mass?
Dr. Sligh: Of course not, you can hold a private mass if you wish, as many as you like, so long as no more than ten people, including yourself, are present.
Fr. Pirnat: Is that your final word?
Dr. Sligh: Yes, Reverend, I'm afraid it is.

Fr. Pirnat went back to his church and decided to hold a private mass to honor the day. Word soon got around that he would be saying such a mass, so on All Saints day there were two hundred people in the church to attend Fr. Pirnat's private mass. Someone called Mayor Hasley, who ordered Police Chief O'Brien to go down to St. Peters and arrest Fr. Pirnat. Chief O'Brien, embarrassed to be arresting a priest, reluctantly informed Fr. Pirnat after the mass that he had violated health regulations and would have to appear before the Police Magistrate Daniels the next day at 1:00 pm. When Pirnat duly appeared the next day in court there was no one there to press charges, so Magistrate Daniels dismissed the case. When the Mayor heard this he was furious and personally completed a formal complaint demanding that Pirnat be made to appear again on Saturday.

By this time the whole town was abuzz with gossip about the incident and on Saturday, large crowds gathered outside the magistrates court in city hall, in contravention of health regulations, of course. The mayor ordered the police to limit the number of people in the building to no more than ten, but the best they could do was to limit those in the court-room to that number. A prolonged acrimonious exchange took place between Mayor Hasley and Magistrate Daniels, the mayor criticizing the magistrate for having dismissed the charges on the previous day and Daniels defending his actions citing the lack of any formal complaint. Fr. Pirnat pleaded guilty to the charge of having conducted a mass on All Saints Day at which more than ten people were present. Daniels, noting Pirnat's prominence and his contribution to the community in building St. Peter's Parochial School, gave him a suspended sentence and cautioned him to strictly abide by the health regulations in the future.[24]

On November 3, Dr. Sligh made up for his previous oversight by issuing a new order that no more than ten people could gather at even an open air affair. But the difficulties in enforcing such strict measures had already become abundantly apparent not only in Fr. Pirnat's church, but in Mayor Hasley's city hall during the trial. The ordinance against congregating was further tested and found wanting a few days later on November 11, when the armistice was announced and an impromptu celebration began at 1:00 in the morning. The Mayor himself violated it by scheduling a formal parade for later in the day. It was no coincidence that the next day the ban on church services and public meetings was abolished and the city announced that theaters and movie houses would soon reopen. On November 28, it was announced that schools would open December 2, the time missed was to be made up during the Christmas vacation time and on Saturdays.[25]

Nevertheless, even while the number of cases was declining, the influenza epidemic continued. On December 8, the Board of Health issued new regulations closing dance halls, limiting funeral processions to near relatives and ordering that placards be placed on all houses where the disease was present. New cases continued to be reported throughout December and into the new year. By Christmas, the number of new cases was down to less than one a day, but there were still death notices.[26]

On December 29, the *Standard* reported that the death toll from flu across the U.S. had reached 300,000. This greatly exceeded official estimate of 112,000 for deaths of U.S. servicemen in the war. These were two scourges that no one wanted to see repeated. Anaconda counted its blessings at being able to limit the number of deaths from influenza to thirty even as it mourned twenty-seven of its young men killed while serving their country.[27]

29. End of an Era

The announcement that the Armistice had been signed and the war ended flashed over the Associated Press wire in Anaconda at one o'clock in the morning of November 11, 1918. Offices at the smelter, the foundry and the BA&P railroad were immediately notified, and the town was awakened by a rising din of smelter whistles, sirens and bells. People, some of them half dressed, rushed from their houses and began congregating in the streets. Fire Chief Collins and his men mounted their big fire truck and drove through town with the siren blaring. Church bells joined in and Frank Provost rounded up members of his band and started a parade up Main Street. More and more people joined in until the there was a solid mass of marchers from curb to curb. The saloons opened their doors and the celebrating began.

Mayor Hasley announced that a more formal march would be held at two o'clock that afternoon, but this didn't dampen the spirits of early morning revelers who continued parading through the streets until well past dawn. The Company had not yet declared a holiday so men on the day shift, with their lunch buckets under their arms, caught the early morning street car at 6:30 am as usual. But this news called for a celebration, the war was over, the boys would soon be coming home. How could they work on a day like this? When the street car reached the smelter many had changed their minds about going to work. They didn't get off the car but stayed on and rode back down the hill with the night shift crews which had just gotten off. All together they joined the revelers in town, and the crowds grew. Coal wagon drivers and delivery men returned their wagons to the barns, unhitched their teams and became a part of the growing throng on the streets. The BA&P engine crews returned their locomotives to the roundhouse and went out to celebrate. One crew just abandoned its engine near the depot and went up Main Street to join in the fun.[1]

The afternoon parade took over an hour to pass, with all the fraternal orders and labor organizations marching. John Marchion of the GAR was grand marshall, followed by Civil War and Spanish-American War veterans. A scattering of soldiers on active duty led the parade. The boy scouts were all in uniform, a Scottish bagpiper in full regalia paraded along, and a number of the groups carried gigantic American flags as well as the banners of their organizations. The Italians waved the red, green and white flags of their country and danced along to the tune of accordions. They were followed by parade constable Tell ("Rabbit") Moore and the Negro contingent who marched ahead of a rickety horse-drawn wagon driven by Nick Sanantonio. It was dragging an empty German helmet behind and had a sign announcing "crowns for sale cheap", alluding to Archduke Franz Ferdinand and the Austrian royal family. Frank Provost urged the men in his band to play as loud as they could, although many of them had been at it, off and on, since two o'clock in the morning. Practically every car and truck in town joined the column bringing up the rear of the parade. The Italians insisted that one day was not enough for such a momentous occasion, so made arrangements for a dance the following evening. Butte was caught up in the same holiday mood and impromptu celebrations went on there for three days.[2]

There was, indeed, much to celebrate. The carnage had ceased, the allies, with help from the United States, had emerged victorious. The United States had taken its place as the world's financial capital and one of the world's leading industrial nations. It had become a power to reckon with. Copper production from Butte and Anaconda contributed to the victory and would be important to the country's continuing growth.

On the local level, the holiday mood quickly turned to one of thanksgiving. Anaconda had much to be grateful for. Of over 1,200 men from the town who had gone off to serve their country, casualties were mercifully small with twenty-seven dead and twenty-six wounded. The boys would soon be home. And even though the flu epidemic was continuing, it was evident that the worst was over and the number of deaths was dropping. The flu quarantine had been lifted and Anaconda had just gone over the top on the fourth and last Liberty Bond drive. The future looked bright with the new smoke stack almost completed.[3]

Thanksgiving day observances, following so soon after the armistice, were especially festive and heartfelt, with bounteous family feasts at home and solemn services at all the city's churches. The Christmas season that followed was not only joyous but marked by record spending. The city's merchants were making up for slow sales during the quarantine. The *Anaconda Standard* put out a special fourteen page supplement for Thanksgiving, its "Peace Edition" with articles honoring America's armed services and Montana's contribution to the war effort. The country and the world had been through a difficult period. Better times were ahead. As if to underline its optimism, on Sunday, three days later, the *Standard* came out with Part II of its Special Edition to dedicate the completion of the new stack. "World's Largest Stack Completed at Washoe Smelter". Anacondans, with gratitude in their hearts, looked to the new year with hope.[4]

But the *Standard's* optimism was unfounded. Armistice in Europe did not produce labor peace in Butte nor in many other places in the United States. The year 1919 was the most strikebound in American history. It is estimated that there were over 3000 strikes and 4 million strikers. In most cases, the issue was wages. Wartime wages had not kept up with wartime inflation. In Seattle, a strike of shipyard workers turned into the first general strike in the nation's history. In the East, 350,000 unorganized steel workers went out en mass, bringing all steel making activities to a halt in fifty towns and cities. This was the first modern strike involving an entire industry. In the Appalachian coal fields, John L. Lewis took 450,000 coal miners out on strike and won a 31 percent increase in base pay, raising it to a minimum of $7.50 a day, the highest wage in the industry.[5]

In Butte, the end of the war meant copper surpluses, and the price of copper quickly fell from twenty-five cents to only twelve cents a pound. Barely a month after the war ended the Company closed its Anaconda and Neversweat mines. By January 1919, fifteen mines were down, idling over 7,000 miners. Then on February 6, the Company announced an across the board wage cut of one dollar a day in Butte, Anaconda and Great Falls. The next day, three thousand miners belonging to the MMWU stopped work. They organized a picket line and effectively closed every mine on the hill. On February 9, troops were brought in. The Butte Army-Navy League at a regular meeting endorsed the strike, which it said was called because of the high cost of living. A formal protest against the use of Federal troops was sent to the Secretary of War. However, there was no unanimity among the many unions and once again the strike fizzled without accomplishing anything. The MMWU formally called off the strike on February 17.[6]

On July 4, the IWW called another strike. Eight unions voted to support them, while twenty-two others voted to stay on the job. Again the strikers soon went back to work. However, there was more trouble brewing. The AFL Metal Trades Council representing nine craft unions in Butte, Anaconda and Great Falls presented a new contract proposal to the Company asking for eight dollars a day, a five and one-half day week, double time for overtime and holidays and recognition of seniority. The other craft unions followed with similar demands. The Company offered a one dollar raise and a two-year contract, but the Metal Trades unions rejected it over the seniority clause. A strike was called on August 12, but again there was no solidarity. Only the Metal Trades unions struck, 600 men in Butte, 350 in Anaconda and 100 in Great Falls. They held out until October 14, then voted to return to work.[7]

Matters turned ugly again in April 1920. The IWW called a strike to protest the continued incarceration of Eugene Debs, Big Bill Haywood, Tom Mooney and other war protestors. The mines were closed completely and a mass meeting was called. Demands were published asking for seven dollars a day, a six-hour day, abolition of the rustling card and an end to contract work. On the third day of the strike, a large crowd congregated on the Anaconda road near the Neversweat mine. The sheriff and his deputies, fearing trouble, tried to disperse the crowd. A shot rang out. Then as the crowd scattered, there was a fusillade of shots in reply from the sheriff's men. Two miners were killed and fifteen people were seriously wounded. Like so much of the violence in Butte, it was never discovered who had started the shooting. Federal troops once again were called in and set up military patrols on all the roads. On May 13, the IWW called off its strike and the Company recalled the rustling cards of all IWW members, effectively denying them future employment.[8]

The Anaconda unions, concerned primarily with steady employment, had avoided most of the turmoil going on in Butte. The minds of the member was focused on other things such as that curiosity called Prohibition, which came in the wake of the Armistice.

The Montana prohibition law, went into effect on January 1, 1919, a full year before the law took effect nationally under the Volstead Act. Well before this the temperance lobby had made serious inroads into the production of alcoholic beverages in the United States. All whiskey manufacturing in the country had been prohibited eighteen months earlier under the provisions of the food control law because it was claimed that 40% of the grain in the U.S. went into producing whiskey. Hard liquor prices had reached $23.00 a gallon for whatever stock still remained as the January 1, deadline approached. The three breweries in Butte ceased their operation early in December, 1918 and the Anaconda brewery followed suit before the end of the month.

This singular event was an early sign that Montana as a state was changing, and that the political center of gravity was shifting away from Butte and Anaconda. By 1918, the state's population justified the creation of two congressional districts, the western mountains and the eastern plains. The former was generally liberal and Democrat, and the latter conservative and Republican. The war and high prices for basic grains had encouraged a continuing wave of homesteaders to settle the vast grasslands of the eastern part of the state, former home of the buffalo and later range for cattle and sheep. Their attitudes toward alcoholic beverages and many other things were quite different from those of the miners, smeltermen and loggers of western Montana, whose votes had previously been decisive on such issues. This time the eastern Montana vote, combined with that of newly enfranchised women, dragged Butte and Anaconda precipitously into a state of affairs which would immediately complicate their lives.

To make matters worse, by late spring of 1920, the prosperity of the war years had disappeared and the country was caught in a serious recession. Nationwide, 4.3 million people, fifteen percent of the work force, were unemployed. Though Butte's copper production in 1919 was only half that of 1918, copper was still a glut on the market. The Company had enjoyed such great prosperity in 1916 and 1917 that it raised its annual dividend from $19.75 a share to $23.50. But this was cut to $8.00 in 1918 and to zero in 1920 and 1921. In February 1921, the Company closed down its entire operation. Once again all the mines in Butte were closed and the furnaces of the smelters in Anaconda and Great Falls went cold. After less than three years in service, the big stack in Anaconda, that symbol of the town's worth, stood cold and stark against the winter sky with no smoke issuing from it.[9]

The results of the shut-down in Butte and Anaconda were devastating. The Company was the sole employer, and in the days before unemployment benefits or other federal or state assistance programs, the complete cessation of all economic activity in a community was a catastrophe of major proportions. The only relief came from the meager savings of individuals and from neighbors and friends who were in similarly tight circumstances themselves. Within two months, reports of serious hardship and hunger were heard and both communities organized as best they could to help the neediest. Local merchants extended credit until their own credit ran out, hoping that any day the crisis would end.

But it didn't. It lingered on month after month for almost a year. Many men left the area looking for work. Entire families packed up their belongings and moved on, to go back to the farm, to stay with relatives or to put faith in the gods that something would turn up somewhere else. Many went to Seattle, Tacoma, Portland and San Francisco, starting small colonies of Butte and Anaconda natives. Some even went back to the "old country". For the first time since these towns were founded, there were more people leaving than moving in. Even some of the old-timers who had helped build them were going elsewhere to gain a livelihood. Some would eventually return. Many would not. It was a time of desperation and despair.

This was the end of an era for these two communities. Their glory years were behind them. Butte, the flashy, boisterous, free-spending mining camp and Anaconda, its younger sister, both had passed the high water marks of their existence. Butte, which had grown steadily over the years, had become a major industrial hub of the Northwest, and by 1914 had a reputation all over the U.S. not only as an important mining center but as a wide-open gambling and show town, an early-days Las Vegas. Silver Bow county, for all practical purposes contiguous with greater Butte, had a population of close to 100,000 by 1917. When the 1920 census was taken Silver Bow county had already registered a significant decrease with only slightly more than 60,000 people counted. The strikes of 1917, '18 and '19 had already started the exodus.

Anaconda, a smaller and more cohesive community, lost some old-time residents to more attractive jobs in West Coast cities during the war, but newcomers had arrived to take their place. However, the smelter would never again employ as many workers as during the war. Employment reached an all-time high in 1916, of just under 5,000 men employed and the total remained above 3,000 through 1918 (See table, page 183). Even in the later peak production years of 1929, 1936, 1942-45, smelter employment never approached that number. An economic report for 1945, for example, states that the average non-farm employment that year in Deer Lodge County was 3,741. This includes the ACM, the BA&P, Montana Power, Intermountain Transportation Company, the hospitals at Warm Springs and Galen, and all the merchants and small businesses in the county.[10]

Neither city went into precipitous decline as a result of the recession of 1921, but both emerged from it with diminished potential for the future. While the smelter opened again in January, 1922, Butte and Anaconda were no longer at the center of the Company's plans. Ryan and Kelley would now concentrate their attention on acquiring new sources of copper in Mexico, Poland and Chile and on vertical integration of operations by expanding in other parts of the United States. The ACM Company was no longer just a Montana operation but a prominent US corporation with growing international holdings. These changes would take a another half-century to play themselves out with neither city recognizing their long-term consequences nor willing to accept a gradual erosion of their importance within the state.

As for dame Anaconda, she had put down strong roots and she still carried a fierce pride. After all, the smelter was still one of the largest in the world and a technological pioneer. It would continue to lead into the indefinite future. In fact, some predicted that even if the Butte mines played out, the smelter would continue to operate. Growing concerns about the environment would surely foreclose opening up a similar facility anywhere else in the United States. Thus, with the smelter operating again, Anaconda continued to see itself as a trend-setter in the Marcus Daly mode. Hadn't it hosted the statewide curling championship bonspiel in 1919, with twenty competing teams? Had not the Anaconda high school football team won the state championship in 1921? Before the 1921 recession, the town had floated a bond issue to build the first concrete highway in the state. It would run from the Montana Hotel east on Park Street along the gravel road to Butte as far as the county line near Gregson where it would meet up with a similar concrete highway to be built from Butte.

The new highway was inaugurated in October, 1920 with the Governor of the state doing the honors. As part of the ceremony, all the automobiles in the county were invited to drive in double file the full length of the road and return. Speeding along at twenty miles an hour they made the round trip in an hour. This was Anaconda's first bow to the growing number of automobiles on its streets, with only Main Street and Third Street having been paved previously. East Park Street, the beginning stretch of the paved highway to Butte, would thus be the third street to have that distinction. With the onset of the recession and slow growth thereafter, they would remain the only paved streets for the next fifteen years. Butte did not complete its portion of the highway until 1923. Anaconda still considered itself a pace-setter.[11]

Nevertheless, Anaconda in middle age had lost the dynamic of her youth. As if to underline this changed status the Company informed the town in 1927 that Marcus Daly's *Anaconda Standard* would be closed and reopened in Butte as the *Montana Standard.* This was the event that brought down the curtain on the Anaconda of Daly's vision. The old race track had burned down in 1911 and its replacement had long since ceased to function. The old company store, the proud Copper City Commercial Company, which had boasted being the largest retail establishment in the state, closed its doors in 1921, a victim of the prolonged shut-down of the smelter. The BA&P extension to Georgetown Lake and Southern Cross was terminated in 1925. The Montana Hotel, refurbished in 1903, operated for the next twenty years as a busy hostelry for visiting Company officials from Butte and New York, actors with the many vaudeville companies who played the local theaters and traveling salesmen. But now, it, too, was going through a difficult transition. Motion pictures were supplanting traveling stage groups, and the automobile and the new highway between Butte and Anaconda made it possible for Company officials to drive to town, transact their business and return to Butte without staying overnight. In spite of general prosperity after 1922, business at the hotel was off.

It was the Montana Hotel and the *Anaconda Standard* which had given the town status and class in the state. In 1931 when *Time* Magazine wrote a belated obituary on the closing of the *Standard*, mentioning the paper's earlier days of glory, it stated that at the turn of the century the *Standard* was one of the best edited dailies in the United States. The status derived from this and similar accolades the paper had received formed an important part of Anaconda's image of itself. While the town clung tenaciously to those memories, its image was changing.

But the loss of the *Standard* was not the only event which changed the town that Daly built. It suffered another serious blow in 1929, when the old Margaret Theatre burned down. The building had been completely remodeled and rechristened the "Sundial" in 1927. After only two years in operation it caught on fire and burned to the ground.

Another small indicator of change was the inauguration of the first radio station in the state in 1922. It was not in Butte, as one might have expected, but in Great Falls. Butte got its own station in 1929, Anaconda not until 1955.

But the town's spirit and pride sustained her through the depression of the 1930s, World War II and into the post-war years. But during that latter time she suffered further indignity and erosion of her turn-of-the-century charm as a result of ill-conceived urban renewal projects and the loss of defining landmarks such as St. Paul's church, the Beaudry Hotel, St. Peter's school, the Anaconda High School and Marcus Daly's house. Bowing to the unfathomable dictates of progress, buses replaced the old street-car line. The Montana Hotel closed its doors and a few years later was decapitated, leaving only its skeleton standing. But the crowning blow came in 1980, when Atlantic Richfield, which had bought the Anaconda Company, closed the smelter. This ended for all time the enterprise which had brought the town into being and provided its livelihood.

Already in 1920 signs pointed to an end of an era of promise, but they were in no way so definite as the events of the late 1970s and 1980, which closed the book on Anaconda as an industrial city.

30. Melting Pot Magic

Another sign that times had changed in the Smelter City was the end of large-scale immigration. Jobs on the "hill" were harder to come by. Since large families were the norm among most of the earlier immigrants, with many having from five to ten children, local residents and their children as they came of age could easily satisfy the Company's needs. The town had probably reached a temporary population peak in 1900 when the Washoe Smelter was being built, but it continued absorbing new immigrants into expanding smelter operations through 1920. However, that was the end of the large-scale influx of foreign born into the town. By then nationality patterns had been set and would remain the same for the rest of the century. Except that American born offspring occupied an ever larger proportion of the whole, the community's ethnic mix became frozen in time.

A 1913 Anaconda Company payroll, which lists individuals by place of birth, is a good indicator of the town's various nationalities and what the pattern would continue to be from then on. At the time, foreign born outnumbered American born about two to one, but already, the American-born sons of earlier immigrants were filling more and more jobs to the point that American born had become the largest single group. They were followed by the Irish, the Austrians (Croatians, Slovenians), the Scandinavians (Norwegian, Swedes, Danes), the British (English, Scotch, Welsh, Isle of Man, Canadians), and the Italians, in that order. There was also a small scattering of other nationalities such as Serbs, Germans, French, Swiss and Belgian.[1]

The Jewish and African-American communities were minuscule and the Chinese had all but disappeared. A handful of Jewish merchants, such as Ben Falk and David Cohen, set up shop in the early days of the town and were succeeded in later years by a few more, such as Jacob Schwartz and Sigmund Goodfriend. They became business leaders and civic activists. However, there were never enough of them in town to build a synagogue, so they traveled to Butte for religious services.

The first African-Americans came to Anaconda in 1889 when Marcus Daly brought in twelve to work as waiters and porters in his new Montana Hotel and a few years later two or three more to work as porters in his private railroad car. They became the core of a small black community which settled in the few blocks north of the railroad tracks on Pennsylvania Avenue.

In 1918, the local draft board reported that of the 2,768 men who had registered for the draft up to September of that year, 1,509 were native born, 143 naturalized, and 1,116 were aliens. Of the total, there were only thirty-five blacks and two orientals.[2]

It was not only the changed economic circumstance of limited employment opportunity at the smelter which slowed the flow of foreign-born to Anaconda but a more general change in the nation's attitude toward immigrants. Congress, in a series of immigration acts from 1917 to 1924 imposed ever more restrictive requirements on foreigners wanting to migrate to this country. The Immigration Act of 1921 limited the number of new migrants from Europe, Australasia, the Near East and Africa to three percent of that nationality residing in the U.S. in 1910. The Act of 1924 was even more restrictive, reducing annual quotas to two percent of those nationals recorded by the 1890 census. The effect of this was not only to cut the total numbers of new immigrants to the U.S., but to favor the immigration of Northern Europeans and discriminate against those from Eastern and Southern Europe.[3]

Because Anaconda was a small town, and almost everyone worked for one employer, it had never divided into strictly ethnic neighborhoods. The only really noticeable division was based on income. The more affluent, company superintendents and some prosperous business people, lived on the south side, west of main street, where Marcus Daly had built his house. The poorer neighborhoods were on the east side. Of the two Catholic congregations, St. Pauls, with its primary school located west of Main St., was considered tonier than St. Peters on Fifth and Alder. But even these divisions were relative because most of the town consisted of working class families who lived from payday to payday. Even on the west side, the large majority caught the street car on Third Street every day with lunch buckets under their arms. As a result ethnic and cultural differences did not loom large in the everyday life of the town and there was a great deal of cultural interchange.

To be sure, there were forces of separateness dividing the nationality groups. Some few who arrived had never intended to stay. Their goal was to work hard, make a nest-egg and return to the old country. These were often resented because they were not contributing to the growth and health of the community. This charge, one of several leveled at the Chinese around the turn of the century, was rather hypocritical since the law specifically denied them citizenship. But there were some among all the nationality groups who openly boasted about their expected return to their native land.

Also, there was a natural clannishness in all the immigrant groups expressed by the formation of their own social, fraternal and church organizations and in their efforts to help new arrivals of the same nationality. Old hatreds from Europe also persisted—Irish memories of English oppression, ancient conflicts between the Croats and the Serbs, the less volatile but remembered differences among Swedes, Norwegians and Danes, French-Canadian resentment of English dominance, etc. Inter-marriage between ethnic groups was discouraged by most who were born in the "old country".

So beliefs and prejudices brought from Europe continued to influence family life and individual choices. Added to this there was a feeling of superiority among the early settlers and town founders over the more recent arrivals; and among those who spoke English over those who didn't. Hence while there was contact and association through a common place of work and working conditions, walls still separated the nationality groups and preferences within each helped maintain its separateness.

But the homogenizing magic of the melting pot had already eroded many of these factors by the end of the World War I. The curtailing of immigration in the twenties and the aging of earlier immigrants made differences among the nationality groups seem even less important. Most significantly, the first generation of native born sons and daughters absorbed the atmosphere around them and found they were very much like their school-mates, friends and neighbors from differing ethnic backgrounds. They resisted speaking the language of their parents. They were Americans and wanted to speak American. Some parents encouraged this attitude. Others tried to keep the native language alive.

Enrollment in both the public and parochial schools reflected the community and was therefore a broad ethnic mix which produced further integration, cross cultural exchange and a narrowing of differences. This was received by the older generation with mixed emotions. While recognizing the advantages of being an integral part of the new culture, they lamented the loss of understanding of the old ways. Perhaps the most upsetting for many was the marriage of their offspring outside their own nationality group, but by the 1920s this was becoming more common if not yet readily accepted.

Also, many of the new generation had little interest in joining and supporting the ethnic fraternal and social organizations which had been the center of life for their parents. As a result many of these organizations ceased to exist. Others held on but had greatly reduced memberships. The leaders of these organizations condemned the attitudes of the young and their lack of interest in their heritage.

A striking example were the Irish societies in Butte and their counterparts in Anaconda. Up to the time of the war, these societies were some of the most reliable sources in the United States of continuing contributions for every complexion of Irish cause. Resistance to English oppression and the financing of rebellion was always a top priority for them. It was difficult for them to muffle their outrage when the U.S. joined Britain in the war against Germany. And it was equally hard for them to face the fact that American-born Irish had no interest in what happened on "The Old Sod". Even Gaelic football which long enjoyed a large following and thrived even into the twenties finally lost its popularity. How could the Irish societies keep the spirit alive among young people when they were losing interest in Irish sports and wouldn't join the societies, despite the urging of their elders?

They turned to the church. Their contributions still meant something to the church. So they prevailed upon the Bishop to require that Irish history be taught in parochial high-schools throughout the diocese so that the younger generation would remember. As laudable as these intentions might have been, it seemed a bit incongruous in St. Peters High School in Anaconda for descendants of Croatians, Slovenians, Italians, French, German, and Belgians learning about Brian Barou, the battle of the Boyne, and the Easter Uprising and getting this special dose of Irish history intended primarily for the thirty or forty percent of the class with Irish roots. On the other hand, it offered a welcome contrast to the Italian and French bias prevalent among many of the nuns. Of course, those Irish descendants who went to public schools remained blissfully ignorant of their glorious past.[4]

Perhaps the most striking example of melting-pot magic was the birth of a unique Butte-Anaconda phenomenon, Bohunkus Day. It was started by Butte High School students in the twenties and adopted immediately in Anaconda. This was a student-organized, student-run outlandish dress parade which took place every spring sometime between April 1 and May 1. Classes were suspended and all high school students were expected to march in bizarre costumes through the business district accompanied by the high school band, similarly attired. To march or not to march was not an option for the students, it was an obligation. Enforcers were appointed and anyone not in proper attire was hauled before a student organized kangaroo court, where appropriate punishment was meted out. To most, of course, it was a day of great fun, a wonderful opportunity to dress up and do crazy things. A day to look forward to. Hence there were few recalcitrants for the court and its enforcers to pursue.

While this was strictly a student affair, the school authorities permitted it and it came to be a community event everyone accepted as a day for fun and nonsense. This curious activity can be viewed in many ways, but there is no doubt that it was a wonderful sublimation of ethnic differences. The name "Bohunkus" points definitively in that direction and harks back to the so-called Bohunk scare in Butte in the early 1900s when the largest contingents of southern Europeans were arriving. Yet, here were students in the 1920s, descendants of protagonists on both sides of that episode, making fun of it and getting their elders to sanction the fun and ridicule ancient fears. The intent of the men who first used the term "Bohunk" and the actions they advocated may not have been so benign. But community participation converted Bohunkus Day into a marvelous example of melting-pot magic which continued for decades. And so it was with other celebrations. St. Patrick's day, of course, was a wonderful excuse for everyone to become Irish for the day.

Another example of this sharing of cultural holidays was the Croatian celebration of the Mesopust. As observed in Anaconda and Butte, this was a pre-lenten expiation of the community's sins achieved by heaping them all on the effigy of an ancient and much maligned malefactor called Mesopust who is tried for his sins, found guilty and summarily stabbed to death or hanged. This would take place as the culmination of an elaborate program which included speeches emphasizing ethnic values and toasts to community leaders.

Typical of such observances was one that took place in Butte in 1930. Over one thousand people attended a three-day celebration at Turpin Hall. By this time it had become such an integral part of Butte culture that civic and political leaders, regardless of nationality, were invited and attended gladly. Prominent among the guests at this celebration were sheriff Angus McCloud, ex-sheriff Larry Dugan and Judge Jeremiah Lynch. Even the main speaker bore the strikingly non-Croatian name of Thomas Walker. A similar event in Anaconda in 1937 had a comparable roster: Mayor Thomas McCarvel, Judge R.E. McHugh, County Attorney W.R. Taylor, Police Magistrate Hubert LeJeune and Dr. T.J. Kargacin.[5]

From all these influences and continuing intermarriage in the following years nationality lines were blurred, old prejudices diminished and a new loyalty was formed for the United States and for Dame Anaconda as she settled into middle age. The battles and hatreds of the Old World had no place in this scene.

The only people in the community for whom the melting pot didn't work and who suffered outright discrimination were the blacks. Never a very large community, but of sufficient numbers to have their own church and social club, they were not served in the town's bars, hotels and restaurants and were relegated to the back seats of the balcony in movie houses. When the Company first began hiring blacks for work on the smelter, they were restricted to work on the trams in the roaster hauling calcine, one of the dirtiest and most dangerous places on the "hill", or assigned jobs as janitors in the general office. Though their children went to the local public schools and were active in school athletic programs, socializing outside the classroom and sports was minimal. And from time to time there were isolated acts of racial violence against them. In this, as with immigration, the town pretty much reflected the temper of the times throughout the country.

Thus, by 1920, Dame Anaconda had settled into what became her century-long identity fashioned during her glory years. Her population was predominantly American born and had benefitted from the melting pot phenomenon that sublimated most ethnic differences and conferred on disparate nationalities a new and common heritage and loyalty. Anaconda was not greatly affected by the migratory waves, both internal and foreign, which came in later years and radically changed the character of many towns and cities across the country. While this, of course, reflects a high degree of stability, it also reveals an absence of the dynamism which had characterized her glory years.

And it was precisely this factor which signaled probably the most profound change of all. In her formative years, Anaconda was a magnet that attracted people to settle and work there. But after 1920, there was a slow but continuing flow in the opposite direction. Not only did the town attract few new arrivals, but it could not even provide employment to many of those born and raised there. It had only limited opportunities to offer now for a limited population. Many of its young people had to go elsewhere to earn a livelihood.

There were a few brief periods when the work force on the smelter temporarily increased. One of these occurred in 1929 when the town registered its all-time high in the national census of 12,494. Then in the 1930s the population returned to the lower 1920 level where it remained until the late 1960s when it began a slow decline. Anaconda became a town more people came from than lived there. The account of those years and how the descendants of the settlers of this proud town faced up to the final closing of the smelter in 1980 is a story for another day.[6]

NOTES

Chapter 1. "Mike, We've Got It!"
1. Isaac Marcosson, *Anaconda*, 31.
2. Michael P. Malone, *The Battle for Butte*, 25.
3. Malone, 25.
4. Marcosson, 46
5. Malone, 28. Marcosson, 44.
6. Samuel Eliot Morison and Henry Steele Commager, *Growth of the American Republic*, 86. Paul Rodman, *Mining Frontiers*, 85.
7. Robert Louis Stevenson "Across the Plains", 50-52, as cited in Morison 107. Morison 86.
8. Alvin M. Josephy, *The Nez Perces Indians and the Opening of the Northwest*, 404. Rodman, 85.
9. Marcosson, 16,17.
10. Spense, Clark C., *Territorial Politics and Government in Montana, 1864-1869*, 21.
11. Marcosson, 18. John K. Hutchins, *One Man's Montana*, 125.
12. Marcosson, 18.
13. Malone, 7. Marcosson, 20, 21.
14. Marcosson, 20, 21. Malone, 8.
15. Marcosson, 8. Malone, 22-25. Mary Dolan, *Anaconda Memorabilia 1883-1983*, 2.
16. Marcosson, 18-25.
17. Marcosson, 41-43. See also Minar H. Shoebotham, *Anaconda, Life of Marcus Daly the Copper King*.
18. Conrad Kohrs, An Autobiography, 19. Federal Writers Project, *Montana, a State Guidebook*, 46, 48. K. Ross Toole, *Montana, An Uncommon Land*, 110.
19. Josephy, 63. RHPP Appendix B Historic Context July 12, 1993.

20. Hal Waldrup, manuscript "The Teepee of the Deer", 3, cites Reuben Thwaites, *Early Western Travels*, Vol. 27.
21. Josephy, 302.
22. David Lavender, *The Rockies*, 167.
23. Lavender, 168.
24. Lavender, 141. Prospectus booklet Allen Gold Mining Company, Report on French Gulch. George Tower Jr., Extract from 8th annual State Bureau of Labor, Agriculture and Industry, Kelly Archive, Hearst Free Library, Anaconda, Montana.
25. Robert D. Oakley, U.S. Forest Service, *The Philipsburg Story*, 8-10.
26. Oakley, 7.
27. Roadside Historical Marker, Montana Highway 1, 10 miles west of Anaconda. Kevin Heaney, *Ghost Town Quarterly*, Summer, 1988, 24.
28. Lavender, 171.
29. Lavender, 167. Kohrs, 19.
30. Burlingame & Toole, History of Montana, Vol II, 11.
31. *Anaconda Standard*, June 17, 1906. (hereafter cited *Anaconda Std*)
32. *Anaconda Std.*, Aug 8, 1935.
33. Kohrs, 19.
34. Lavender, 185. Kohrs, 18-21.

Chapter 2. Heaven On Earth
1. For more on Morrisites see LeRoy C. Anderson's *For Christ Will Come Tomorrow*.
2. A.C.M. Records, Anaconda Deer Lodge County Historical Society RB 11, FILE 16.

3. Deer Lodge County History Group, *In the Shadow of Mt. Haggin*, 134. *Anaconda Std.* Aug 8, 1935.

4. Anderson, 58.

5. Anderson, 60.

6. *Journals of Granville Stuart*, 177, as cited by Waldrup, "The Teepee of the Deer", 2. Also see *In the Shadow of Mt. Haggin*, 1; and Meidl & Wellcome, *Anaconda, Montana A Century of History*, 35.

7. Waldrup, 2.

8. Waldrup, 3.

9. *Montana Std.*, Nov. 23, 1955. Waldrup 3. *Montana Std.* Dec 5, 1993.

10. *In the Shadow of Mt. Haggin*, 175.

11. *In the Shadow of Mt. Haggin*, 81, 84, 99.

12. Meidl & Wellcome, 30.

13. *In the Shadow of Mt. Haggin*, 97a, Meidl & Wellcome 50.

14. *In the Shadow of Mt. Haggin*, 129.

15. Meidl & Wellcome, 50.

16. *In the Shadow of Mt. Haggin*, 86. Meidl & Wellcome 16.

17. *In the Shadow of Mt. Haggin*, 77.

18. Meidl & Wellcome, 35.

19. *In the Shadow of Mt. Haggin*, 21.

20. Meidl & Wellcome, 34.

21. Matt J. Kelly, *Anaconda, Montana's Copper City*, 48.

22. *In the Shadow of Mt. Haggin*, 21.

23. *Montana Std.*, Dec 15, 1941.

24. Kelly, 49.

25. Kelly Collection Scrapbook, 48.

Chapter 3. Boom Town

1. Marcosson, 27, 80.

2. Marcosson, 47.

3. Minar H. Shoebotham, 74.

4. Marcosson, 35.

5. Kelly, 9.

6. Lavender, 261.

7. *Anaconda Std.*, Sept. 24, 1947. Kelly, 10. W.E. Bowden 1977 speech, A.C.M. records file 41, box 5, General Office file.

8. Bob Vine, *Anaconda Memories*, 2. Kelly, 10, 11. A.C.M. Records RB 11, file 16.

9. Kelly, 11.

10. Marcosson, 261. Kelly, 12.

11. Kelly, 12.

12. Lavender, 279.

13. Vine, 4. Kelly, 48. *In The Shadow of Mt. Haggin*, 168.

14. Kelly Collection, EMC 48.

15. Vine, 4. Kelly, 12.

16. Kelly, 12.

17. Vine, 3.

18. Anaconda section of Butte City Directory 1885, 1886.

19. Kelly, 62. *Montana Std.*, June 30, 1933.

20. Kelly, 10. Anaconda Std., May 17, 1945.

21. Vine, 3. *Montana Std.*, Feb. 11, 1934. *Anaconda Std.* Oct 22, 1938.

22. Article by John McNay in special anniversary edition of the *Anaconda Std.* June 25, 1983. *In The Shadow of Mt. Haggin*, 115. *Anaconda Std.*, Sept. 12, 1939.

23. *In the Shadow of Mt. Haggin*, 31.

24. Information on Judge Winston and Bridget Sullivan comes from separate biographical cards on each of them prepared by the Anaconda Deer Lodge County Historical Society for the Anaconda Centennial in 1983. Alice Finnegan, a local historian, tells me that there were two Bridget Sullivans living in Anaconda

and the one working for Judge Winston was not the Bridget of Lizzie Borden fame.

25. Dolan, 44. *Anaconda Std.* Jan. 20, 1901.

26. *Anaconda Std.* Jan. 20, 1956. Anaconda section of Butte City Directory, 1885.

27. Kelly, 16. Dolan, 41. *In the Shadow of Mt. Haggin*, 10, 11, 220.

28. Kelly, 20. Dolan insert. *In the Shadow of Mt. Haggin*, 176.

Chapter 4. 100 Years of Ore

1. *In the Shadow of Mt. Haggin*, 17. Kelly, 12.

2. *Butte Miner*, July 25, 1884.

3. *In the Shadow of Mt. Haggin*, 62.

4. Vine, 4. Marcosson, 58.

5. Vine, 3. Marcosson, 50, 51.

6. Paul Rodman, *Mining Frontiers of the Far West*, 70, 141.

7. Malone, 31., Kelly, 14.

8. Kelly, 15.

9. Marcosson, 52. Vine, 5.

10. Vine, 5. Kelly, 15.

11. *Montana Std.*, Feb 11, 1934.

12. Dolan, 45. Also see *In the Shadow of Mt. Haggin*, 197-217.

13. Kelly, 29.

14. Kelly, 20-22.

15. Vine, 6.

16. Kelly, 18.

17. *In the Shadow of Mt. Haggin*, 222.

18. Kelly, 37.

19. *Anaconda Std.*, Feb. 12, 1942.

20. Kelly, 37-39. *In the Shadow of Mt. Haggin*, 40-41. Dolan, 12-13.

21. Kelly, 29. *In the Shadow of Mt. Haggin*, 181-182.

22. Kelly, 23.

23. Kelly, 23.

24. Kelly, 31.

Chapter 5. Bad Medicine Wagon

1. Marcosson, 14. Malone, 7, 10.

2. Morison, 82. Toole, 124.

3. Toole, 125. Montana, A State Guidebook, 48.

4. Morison, 105-112.

5. William Kitridge, Annick Smith, *The Last Best Place*, 354-364.

6. Dee Alexander Brown, *Hear That Lonesome Whistle Blow*, 81, 85.

7. Brown, 64, 79.

8. Brown, 87.

9. Marcosson, 27.

10. Brown, 206-210.

11. Toole, 128.

12. FWP *Montana* Guidebook, 91.

13. Brown, 256.

14. *In the Shadow of Mt. Haggin*, 171.

15. Maury Klein, *Union Pacific*, 555-558.

16. Klein, 557.

17. Brown, 235-247.

Chapter 6. *"Help, Help, Come Running!"*

1. Merrill D. Beal, *I Will Fight No More Forever*, 48-56.

2. Beal, 65.

3. Beal, 87.

4. Beal, 37-43.

5. Robert I. Burns, *The Jesuits and the Indian Wars of the Northwest*, 431.

6. Burns, 436. Josephy, 566. Beal, 107.

7. Burns, 438.

8. Burns, 425.

9. Josephy, 557.

10. Josephy, 574.

11. Burns, 447. Beal, 112.

12. Burns, 450.

13. Josephy, 577. Beal, 127.

14. Mark H. Brown, *The Flight of the Nez Perce*, 260.

15. Mark Brown, 265.
16. Mark Brown, 279.
17. Mark Brown, 276, 282.
18. Beal, 185-276.
19. Walter D. Johnson, *On the North Side*, 18.
20. Toole, 13

Chapter 7. Clark, Daly and the Anaconda Std.
1. Clark C. Spense, *Territorial Politics and Government in Montana 1864,-89*, 10.
2. Spense, *Territorial Politics*, 11-15.
3. Spense, *Territorial Politics*, 23.
4. Federal Writers Project, 161. Toole, 173.
5. Toole, 173.
6. Malone, 84.
7. Morison, 229. Malone, 87.
8. Lavender, 284.
9. Malone, 86. Also see Burlingame & Toole, *History of Montana*, Vol. I. Chapter IX.
10. Emmons, *The Butte Irish*, 102.
11. Shoebotham, 180. Malone, 84.
12. Malone, 87.
13. C.B. Glasscock, *The War of the Copper Kings*, 113. Mary Dolan, "The Role of Newspapers in Anaconda's History", 2.
14. *Anaconda Std.*, Sept. 24, 1889.
15. *Anaconda Std.*, Oct. 17, 1889.
16. *Anaconda Std.*, Sept. 4, 1889.
17. *Anaconda Std.*, June 23, 1891.
18. *Anaconda Std.*, April 23, 1891.
19. Dolan, 2.
20. Shoebotham, 207.
21. Kelly, 20. Dolan, 2.
22. Malone, 90.

Chapter 8. Winning
1. Shoebotham, 163.
2. For more on Daly's race horses see Shoebotham 147-169 and Copper Camp, 230-234.
3. Kelly, 27.
4. *Anaconda Std.*, Jan 16, and 19, 1891.
5. Kelly, 22.
6. Kelly, 50.
7. Kelly, 23.
8. *Montana Std.*, June 17, 1931.
9. *Montana Std.*, June 21, 1931.
10. *Montana Std.*, June 21, 1931.
11. Copper Camp, 226.
12. *Anaconda Std.*, Oct. 26, 1935.
13. *Anaconda Std.*, March 23, and 30, and April 12, 1890.
14. *Anaconda Std.*, April 2, 1890.
15. *In the Shadow of Mt. Haggin*, 61.
16. *Anaconda Std.*, May 16, and June 7, 1906.
17. *Anaconda Std.*, Aug. 10, and 11, 1890.
18. Copper Camp, 59,60.
19. Dolan, 30.
20. *In the Shadow of Mt. Haggin*, 67. Kelly, 50. Dolan, 30.
21. Dolan, 30. *Anaconda Std.*, Dec 14, 1930.

Chapter 9. Growing Up
1. Alexis De Tocqueville, *Democracy In America*, Vol. II, 128.
2. *In the Shadow of Mt. Haggin*, 148. Kelly, 18.
3. *In the Shadow of Mt. Haggin*, 152. Kelly, 18.
4. *In the Shadow of Mt. Haggin*, 153. Kelly, 18. Anacoda City Directory, 1889.
5. David Emmons, *The Butte Irish*, 313.
6. Kelly, 26.

7. *Anaconda Std.*, Aug. 10, 1890.

8. *In the Shadow of Mt. Haggin*, 221, 223.

9. Malone, 56.

10. Kelly, 22. Dolan, 20.

11. Kelly, 22.

12. Kelly, 52. *Anaconda Std.*, Feb. 17, 1893.

13. *Anaconda Std.*, Apr. 27, 1890. Kelly, 25.

14. *In the Shadow of Mt. Haggin*, 56, 57.

15. Kelly 26. Kelly Collection 4A p.34. Anaconda section of Butte City Directory 1886.

16. Anaconda City Directory, 1896.

17. Kelly Collection, 4A p.34.

18. Dolan, 35.

19. *In the Shadow of Mt. Haggin*, 167. Meidl & Wellcome, 40.

20. Kelly, 35. Meidl/Wellcome, 59.

21. Kelly, 35. Meidl/Wellcome, 32.

22. Dolan, 35.

23. Others connected with the store were: George Hutchens, Carl Wadell, Joe Parker, A. Nozell, Art Nicely, Jake Keene, Jimmy White, George Remington, John MacKenzie, Jock Anderson, Lew Coleman, Frank Clinton, Jim Driscoll and Tom Murphy.

Chapter 10. Chinese Boycott

1. See Betty Lee Sung, The Story of the Chinese in America, Chapters 3 and 4.

2. Stacy A. Flaherty, "Boycott in Butte: Organized Labor and the Chinese Community, 1896-1897", 36; *Montana Magazine of Western History*, Winter 1987. Robert R. Swartout Jr., "From Kwangtung to the Big Sky: The Chinese Experiencein Frontier Montana", 44, 45; *Montana Magazine of Western History*, Winter 1988.

3. Swartout, 44, 46, 47. *Anaconda Std.*, undated clipping, May 23, 1917.

4. *Anaconda Std.*, Sept. 18, 1889.

5. Sung, 25.

6. Copper Camp, 110.

7. 1866 undated newspaper clipping. Alice Finnegan file.

8. *Helena Daily Herald*, Dec. 12, 1871.

9. Clipping from *Deer Lodge Montanian*, Sept. 4, 1871, Alice Finnegan file.

10. *Deer Lodge Montanian*, Feb. 1, 1891. Morison, 155.

11. Sung, 44.

12. *Anaconda Std.*, Aug. 5, 1891; and June 2, 1802.

13. Copper Camp, 208.

14. Kelly, 27.

15. Matt Kelly Scrapbook clipping *Anaconda Std.* Nov. (no date) 1891, Jan 1, 1892.

16. *Anaconda Std.*, Jan. 8, 1893.

17. *Anaconda Std.*, Sept. 18, 1889.

18. *Anaconda Std.*, Feb. 17, 1893. Kelly Scrapbook clipping *Montana Std.*, April 22, 1956.

19. Copper Camp, 112.

20. *Anaconda Std.*, Apr. 13, 1892.

21. Sung, 56.

22. *Helena Daily Herald*, Jan. 24, 1870.

23. Report on newspaper opinions in *Anaconda Std.*, May 19, 1893.

24. *Anaconda Std.* summary of newspaper comment, May 17, 1893. Article May 22, 1893.

25. *Anaconda Std.*, Feb. 17, 23, 24, and 25, 1893.

26. *Anaconda Std.*, Feb. 13, 14, 1893.

27. *Anaconda Std.*, Feb. 17, and May 17, 1893.

28. *Anaconda Std.*, Feb. 27, 1893 and March 2, 1892.

29. *Anaconda Std.*, Jan. 11, Feb. 26, and June 6, 1893. Also March 8, 1892.

30. Flaherty, 36-47.

31. Report of Montana State Bureau of Agriculture, Labor and Industry, Helena, 1902.

32. *Butte Miner*, Jan. 3, 1901.

33. *Anaconda Std.*, May 20, 1906

34. *Anaconda Std.*, May 20, 21, 1906.

Chapter 11. A Small Company Town

1. See also Burlingame and Toole, Vol. I Chapter X and Malone 93-98.

2. Kelly, 41.

3. *Anaconda Std.*, Oct. 23, 1892. John McNay, anniversary edition of *Anaconda Std.*, June 23, 1983. Kelly, 41.

4. Malone, 103.

5. Burlingame and Toole, Vol. 1, Chapter X. Malone, 99-103.

6. Malone, 99.

7. Malone, 100.

8. *Harpers Magazine*, July, 1984. Emmons, 99.

9. Lawrence McCaffrey, *The Irish Diaspora in America*, 102.

10. Kelly, 41.

11. Malone, 104.

12. Shoebotham, 102.

13. Malone, 42. Marcosson, 55.

14. Dolan, 8.

15. Kelly, 29.

Chapter 12. A Uniform Wage Underground

1. Morison, 161.

2. Paul W. Rodman, *Mining Frontiers of the Far West*, 69.

3. Malone, 76. Lavender, 308.

4. Malone, 76.

5. Malone, 76.

6. Maury Klein, *Union Pacific*, 490.

7. Morison, 156.

8. As quoted in the FWP *Montana Guidebook*, 70

9. Kelly file, unnumbered. Polk Directory.

10. Kelly File newspaper clipping, *Anaconda Std.* Sept. 8, 1894.

11. Lavender, 309.

12. Lavender, 309, 310. *Anaconda Std.*, Apr. 3, 1892.

13. Lavender, 311. *Anaconda Std.*, May 21, 1893.

Chapter 13. Knowledge and Culture

1. Morison, 250. Lavender, 331.

2. Malone, 55.

3. Anaconda City Directory, 1899.

4. Dolan, 15.

5. W. A. Swanberg, *Citizen Hearst*, 88.

6. *In the Shadow of Mt. Haggin*, 180.

7. Dolan, 14-16.

8. Kelly, 46.

9. Dolan, 11.

10. Kelly, 39. Meidl & Wellcome, 4.

11. *Anaconda Std.*, May 23, 1917.

12. Dolan, 10.

13. *Anaconda Std.*, March 15, 1890.

14. Selected from advertisements in the *Anaconda Std.*, from January to June 1891.

15. *Anaconda Std.*, May 23, 1917.

16. Vine, 15. Kelly, 40.

17. *Anaconda Std.*, Feb. 15, 1891; May 6, 1906; July 8, 1906.

Chapter 14. Free Silver and War with Spain

1. Morison, 137.

2. As early as 1883, The Powder River Cattle Company, owned by

British and New York investors was running from 4,000 to 8,000 cattle, employing over 200 cowboys in eastern Montana and Wyoming.
3. Morison, 238.
4. Morison, 241.
5. Malone, 93.
6. Draft monograph, J.J. Morris, 1938.
7. Malone, 94.
8. Morison, 247.
9. Lavender, 312.
10. *Anaconda Std.*, Jan. 8, 1893. Morison, 156.
11. Jeanette Prodgers, Butte-Anaconda Almanac, 49. Morison, 162-164.
12. Vine, 13.
13. Malone, 107.
14. Malone, 109. Morison, 254.
15. Kely, 43.
16. Kelly, 43. FWP *Montana* Guide, 54.
17. Kelly, 44.

Chapter 15. A New Century
1. Patrick Renshaw, *The Wobblies*, 48.
2. Renshaw, 48.
3. Burlingame and Toole, Vol. I, Kelly, 44. Roberta C. Cheney, *Names on The Face of Montana*.
4. Kelly, 18. *Anaconda Std.*, Sept. 12, 1945.
5. *Montana Std.*, July 19, 1931.
6. Toole, 172.
7. Marcosson, 81-82.
8. Marcosson, 89-90, 92.
9. Marcosson, 95. Toole, 195.
10. Toole, 195.

Chapter 16. *"What's the Price of a Vote Today?"*
1. Marcosson, 114. Malone, 49-51.
2. Marcosson, 117, Lavender, 260, 326.

3. Marcosson, 111-135. Malone, 144-148.
4. Malone, 148-151.
5. FWP *Montana* Guide, 69.
6. Malone, 112.
7. Copper Camp, 126.
8. Malone, 115.
9. Malone, 117.
10. Malone, 121.
11. Malone, 121.
12. Malone, 122.
13. Malone, 123.
14. Malone, 126.
15. Lavender, 327. Malone, 127.
16. Malone, 123, 124.
17. Malone, 153.
18. Malone, 161.
19. Vine, 124. Meidl & Wellcome, 59.

Chapter 17. *"The Mighty Oak Has Fallen"*
1. Marcosson, 97-98.
2. Marcosson, 99.
3. Lavender, 327. Malone, 163, 196.
4. Marcosson, 124.
5. Malone, 173.
6. Joseph Kinsey Howard, *Montana High, Wide, and Handsome*, 119.
7. C.B. Glasscock, *The War of the Copper Kings*, 119.
8. Marcosson, 128.
9. Malone, 193.

Chapter 18. Washoe Smelter
1. Malone, 53.
2. Kelly, 58.
3. *Anaconda Std.*, Oct. 22, 1934.
4. Vine, 19.
5. "A Brief Description of the Anaconda Reduction Works", 1936.
6. Emil Kramlick, 1972 Report, 18.
7. Kramlick, 19. Marcosson, 324.

8. Marcosson, 231.

9. ACM records, 1927. Anaconda Deer Lodge County Historical Society.

10. Interview with Jerry Hansen, Anaconda, Aug. 12, 1995.

11. Taped interview with Martin Judge, Anaconda, Aug. 20, 1995.

12. Anaconda Deer Lodge County Historical Society Taped Interview, Tom Dixon, Nov. 25, 1981.

13. "A Brief Description of the Anaconda Reduction Works"

14. *McClures Magazine*, June 1984, 2-20.

15. Tom Dixon Interview

16. Interview with Bill Flynn, Anaconda, Aug. 31, 1995.

Chapter 19. Coming of Age

1. Anaconda City Directory 1896, and 1899.

2. Vine, 23. Kelly, 54-55.

3. Kelly, 31.

4. Dolan, 44. In the Shadow of Mt. Haggin, 197. Anaconda Deer Lodge County Historical Society 1927 pamphlet History of the High School. *Anaconda Std.*, March 3, 1902, Feb. 27 1903.

5. Kelly, 45. *Anaconda Std.*, June 10, 1937. Evidently the excavation was not properly filled in, but covered with heavy planks and a foot of dirt. Years later the planks rotted and Mrs. Reno Puccinelli plunged 15 feet into a large pit when part of her yard caved in. She suffered a broken arm and leg. *Anaconda Std.*, Sept. 12, 1945.

6. *Anaconda Std.*, Sept. 12, 1945. *In the Shadow of Mt. Haggin*, 199. Dolan, 45.

7. Anaconda City Directory, 1899. There were other slaughter houses not listed which belonged to the owners of the various meat markets.

8. *Anaconda Std.*, Sept. 24, 1936, May 4, 1937, and Nov. 9, 1945. Meidl & Wellcome, 4.

9. Meidl & Wellcome, 4.

10. *Anaconda Std.*, March 18, 1936.

11. Howard, 119.

12. Kelly, 36.

13. Kelly, 42.

14. *Anaconda Std.*, Feb. 17, 1893.

15. *Anaconda Std.*, Sept. 24, 1894.

16. Kelly, 42.

17. Malone, 74.

Chapter 20. Smoke Farmers

1. Marcosson, 105. Meidl & Wellcome, 55.

2. J. McNay, Anniversary issue, *Anaconda Std.*, June 25, 1983.

3. Kelly, 60.

4. Marcosson, 105.

5. Kelly, 60.

6. Anaconda, Deer Lodge County Historical Society Taped Interview, John Holtz, 3/8/84.

7. Interview, Jim Morris, Aug. 28, 1995.

8. Booklet, Deer Lodge County, Montana, 1907. Anaconda Deer Lodge County Historical Society.

9. Morison, 397-400.

10. John S. Baird, *Early Days in the Forest Service*, March 13, 1944, Kelly Collection. McNay, *Anaconda Std.*, June 25, 1983.

11. Glasscock, 234.

12. Malone, 44.

13. *Anaconda Std.*, June 24, 1906.

14. Arthur C. Knight M.D., Superintendent of Galen Sanitarium. Report "Fifty Years of Progress." 1963.

Chapter 21. Unions and Socialists
1. Renshaw, 40.
2. Morison, 139. Garrity, 108.
3. Morison, 145.
4. Renshaw, 59.
5. Jerry Calvert, "The Rise and Fall of Socialism in a Company Town: Anaconda, Montana. 1902-1905." *Montana Magazine of Western History,* Autumn, 1986, 6-7.
6. Calvert, 8, 10.
7. Calvert, 8-9, 12.
8. Kelly, 47. Calvert, 11,12. Jensen, 295, 296, quoting Heinze, *Reveille* newspaper of Sept. 7, 8, 1903.
9. *Anaconda Std.,* Apr. 5, 1904 as quoted by Calvert, 12.
10. Renshaw, 58. Spense, 115.
11. Malone, 77.
12. Renshaw, 62-63.
13. Lavender, 315, 316. Vernon H. Jensen, *Heritage of Conflict,* 72-87.
14. Emmons, 275. Renshaw, 60.
15. Renshaw, 62.
16. Renshaw, 21-26.
17. Renshaw, 26. Patricia N. Limerick, *Legacy of Conquest,* 118, 119.
18. Jensen, 202-203.
19. Jensen, 204.
20. Jensen, 205.
21. Jensen, 205.
22. Jensen, 202-206.
23. Jensen, 206, 207.
24. Jensen, 217.
25. Jensen, 217.
26. Emmons, 275.
27. Malone, 77.

Chapter 22. Silver Jubilee
1. Jensen, 202-203.
2. FWP *Montana* Guide, 70.
3. FWP *Montana* Guide, 71. Jensen, 306.
4. FWP *Montana* Guide, 71.
5. Vine, 23.
6. *Montana Std.,* March 9, 1930.
7. Mathewson was a Canadian. There were a number of Canadians in management at that time. Interview with James McGeever, Anaconda, Aug.10, 1993.
8. *Anaconda Std.,* May 22, 1906.
9. Renshaw, 47.
10. Kelly, 59.
11. Kelly, 59.
12. Kelly, 48. Kelly Collection.
13. *In the Shadow of Mt. Haggin,* 50.
14. Kelly, 49.
15. Kelly, 50. Matt J. Kelly, "Trail Herds of the Big Hole Basin," *The Montana Magazine of History,* July, 1952. *Anaconda Std.,* Dec. 17, 1905, June 17, 1906, Jan. 22, 1913.
16. Kelly, 51.

Chapter 23. John D. Ryan
1. Marcosson, 152. Carrie Johnson, "Electric Power, Copper and John D. Ryan," *Montana The Magazine of Western History,* Autumn 1988, 27.
2. Marcosson, 110.
3. Marcosson, 139. Malone, 205.
4. *Anaconda Std.,* June 13, 1906.
5. Malone, 203.
6. *Anaconda Std.,* June 11, 1906.
7. Marcosson, 144.
8. Marcosson, 145.
9. Johnson, 30-32.
10. Marcosson, 148.
11. FWP *Montana* Guide, 71.
12. Vine, 28.
13. Vine, 26. Marcosson, 147.
14. Malone, 205.

Chapter 24. Expansion and Opportunity
1. Morison, 127.

2. See also Marcella Walter, "Magnificent Distances," 20, *Montana Magazine*, June 1994.

3. *Montana Std.,* July 25, 1993. *Anaconda Std.,* May 16, 1906. Walter, 22.

4. *Anaconda Std.*, May 30, 1906.

5. Bruce Baxter, Montana History Calendar, 1991. Prodgers, 48.

6. Copper Camp, 292.

7. *In the Shadow of Mt. Haggin*, 24-27

8. Marcosson, 110.

9. Marcosson, 153.

10. Marcosson, 285. *A Brief Description of the Anaconda Reduction Works*, 12.

11. Marcosson, 143, 144.

Chapter 25. Wreck the House of the Grafters

1. Proceedings, Twenty Third IUMMSW, 1918, 33-37 as quoted by Vernon Jensen, 380.

2. Jensen, 336, 339.

3. Emmons, 272. Jensen, 232, 234.

4. Jensen, 317.

5. Emmons, 268.

6. Jensen, 318, 326. Emmons, 271-272.

7. *Miners Magazine*, Aug. 15, 1912, 248-250 as cited by Jensen, 316-323. Emmons, 268, 271.

8. Emmons, 272.

9. Jensen, 327.

10. Jensen, 328-329. Copper Camp, 60-65.

11. Copper Camp, 63.

12. Jensen, 333.

13. Emmons, 277.

14. Jensen, 332.

15. Jensen, 334.

16. Jensen, 335.

17. Jensen. 335.

18. Jensen, 344-347.

19. Jensen, 347-349.

20. Jensen, 347, 349. Spense, 124.

21. Jensen, 350.

Chapter 26. War Comes to Anaconda

1. Morison, 445.

2. Morison, 425.

3. Toole, 213-215.

4. Spense, 118.

5. Spense, 121-123.

6. Morison, 471-479.

7. Morison, 475-478. *Anaconda Std.,* Nov. 12, 1917.

8. *Anaconda Std.,* movie play bills in 1917 and 1918.

9. Morison, 472-476.

10. Emmons, 364.

11. *Anaconda Std.*, June 7, 1917. Morison, 481.

12. *Anaconda Std.*, July 13, 1917.

13. *Anaconda Std.,* Sept. 8, 13, 29, Nov. 1, 7, 1917, Nov. 23, 1918.

14. *Anaconda Std.*, Sept. 8, 1917.

15. Morison, 483.

16. Kelly, 62.

Chapter 27. Murder in Butte

1. Emmons, 354.

2. Emmons, 354.

3. Kelly, 60. Marcosson, 155.

4. Emmons, 367.

5. Jensen, 372.

6. Jensen, 376.

7. Jensen, 377.

8. Jensen, 430.

9. Copper Camp, 294. *Anaconda Std.,* July 12, July 14, 1917. FWP *Montana* Guide, 73.

10. FWP *Montana* Guide, 74. For ACM wage offers see *Anaconda Std.*, July 22, and 25, 1917.

11. Jensen, 437.

12. FWP *Montana* Guide, 74. *Anaconda Std.*, July 27, 1917. Jensen, 437.

13. FWP *Montana* Guide, 74. Jensen, 437.

14. *Anaconda Std.,* July 15, quotes article from July issue of Miner Magazine stating IUMMSW declares war on IWW. Jensen, 438.

15. *Anaconda Std.,* Aug. 2, 1917. See also, *Anaconda Std.,* June 11, July 15, 17, 29, 30, Aug. 16, 17, 20, 21, 26, 27, Sept. 8, 9, 11, 17, 1917.

16. *Anaconda Std.,* Aug. 5, 1917. Jensen, 439.

17. FWP *Montana* Guide, 74.

18. *Anaconda Std.,* Aug. 15, 1917.

19. Jensen, 440.

20. FWP *Montana* Guide, 74.

21. Jensen, 442. Emmons, 383.

22. *Anaconda Std.,* July 28, and Aug. 3, 1917.

23. *Anaconda Std.,* Aug. 8, and Aug. 11, 1917.

24. Jensen, 439.

25. *Anaconda Std.,* Aug. 24, and Aug. 25, 1917.

26. *Anaconda Std.,* Aug. 24, 1917.

27. *Anaconda Std.,* Sept. 12, 1917.

28. *Anaconda Std.,* Sept. 4, 1917.

29. *Anaconda Std.,* Sept. 11, 1917.

30. FWP *Montana* Guide, 74. Emmons, 378. *Anaconda Std.,* Sept. 15, and Sept. 16, 1917.

31. *Anaconda Std.,* Nov. 16, 1918.

32. Howard, 89. FWP *Montana* Guide, 75.

33. Geoffrey Perrett, *America in the Twenties,* 57, 94.

Chapter 28. *"Please Do Not Congregate."*

1. Marcosson, 155.

2. Marcosson, 157.

3. Vine, 29.

4. Vine, 29.

5. Kelly, 62.

6. *Anaconda Std.,* Sept. 18, 26, 27, 28, 1918.

7. *Anaconda Std.,* Oct. 6, 1918.

8. *Anaconda Std.,* Oct. 9, 1918.

9. *Anaconda Std.,* Oct. 9, 1918.

10. *Anaconda Std.,* Oct. 11, 1918.

11. *Anaconda Std.,* Oct. 13, 1918.

12. *Anaconda Std.,* Oct. 11, 1918.

13. *Anaconda Std.,* Oct. 13, 1918.

14. *Anaconda Std.,* Oct. 14, 16, 17, 18, 19, 20, 1918.

15. *Anaconda Std.,* Oct. 17, 19, 21, 24, 1918.

16. *Anaconda Std.,* Oct. 17, and 25, 1918.

17. *Anaconda Std.,* Oct. 25, and Nov. 3, 1918.

18. *Anaconda Std.,* Oct. 25, 27, 28, 1918.

19. *Anaconda Std.,* Nov. 1, 1918.

20. *Anaconda Std.,* Nov. 1, 1918.

21. *Anaconda Std.,* Nov. 8, 1918.

22. *Anaconda Std.,* Nov. 28, 29, 30, Dec. 4, 1918.

23. *Anaconda Std.,* Nov. 21, and Dec. 8, 1918.

24. *Anaconda Std.,* Nov. 2, 3, 1918.

25. *Anaconda Std.,* Nov. 3, 12, 28, 1918.

26. *Anaconda Std.,* Dec. 21, 1918.

27. *Anaconda Std.,* Dec. 16, 21, 23, 29, 1918.

Chapter 29. End of an Era

1. Kelly, 63.

2. Kelly, 62, 63. *Anaconda Std.,* Nov. 12, 1917.

3. *Anaconda Std.,* Dec. 16, 27, 1918.

4. *Anaconda Std.,* Nov. 28, Dec. 1, 1918.

5. Perrett, 39, 44, 46, 49.

6. Jensen, 447, 448.

7. Jensen, 449.

8. Howard, 49. Jensen, 450.

9. Perrett, 130. Emmons, 399.

10. *Anaconda Std.*, July 5, 1946.

11. Kelly, 64.

Chapter 30. Magic Melting Pot

1. 1913 ACM payroll, Anaconda Deer Lodge County Historical Society

2. *Anaconda Std.*, Nov 23, 1918.

3. Morison, 188.

4. For fuller treatment of the Irish in Anaconda, see Laurie K. Mercier, "We Are Women Irish," *Montana Magazine of Western History*, Winter, 1994, 28.

5. *Anaconda Std.*, March 2, 1930; and Feb. 9, 1937.

6. U.S. Census, 1930.

BIBLIOGRAPHY

Anderson, C. LeRoy, *For Christ Will Come Tomorrow, The Saga of the Morrisites,* Utah State U. Press, Logan Utah, 1981.

Beal, Merrill D., *I Will Fight No More Forever,* Ballantine Books, New York, 1963.

Brown. Dee Alexander, *Hear That Lonesome Whistle Blow,* Holt, Rinehart and Winston, New York, 1977.

Brown. Mark H., *The Flight of the Nez Perce,* University of Nebraska, Lincoln, 1982.

Burlingame & Toole, *History of Montana,* Vol. I & II, New York 1957.

Burns, Robert Ignatius, S.J., *The Jesuits and the Indian Wars of the Northwest,* University of Idaho, Moscow, Idaho, 1966

Cheney, Roberta, *Names on the Face of Montana,* Missoula, 1983.

Deer Lodge County History Group, *In The Shadow of Mt. Haggin,* 1975

Dolan. Mary, *Anaconda Memorabilia 1883 - 1983,* ACME Press, Missoula, Montana, 1983.

Emmons, David M., *The Butte Irish,* University of Illinois, Chicago, 1989.

Federal Writers' Project, *Montana, A State Guidebook,* The Viking Press, 1939, Somerset Publishers 1973.

Garraty, John A., *The American Nation, The U.S. since 1865,* Harper and Row, New York 1971.

Glasscock, C.B., *The War of the Copper Kings.* Bobbs-Merril Co. Indianapolis, New York, 1935.

Hutchens, John K., *One Man's Montana,* J.B. Lippincott Co. Philadelphia, 1964.

Howard, Joseph Kinsey, *Montana High, Wide, and Handsome,* New Haven, Yale University Press, 1959.

Jensen. Vernon H., *Heritage of Conflict,* Cornell University Press, Ithaca, New York, 1950

_____, *Nonferrous Metals Industry Unionism,* Cornell University Press, Ithaca, New York 1954

Johnson, Walter G., *On the North Side, Philipsburg, MT* 1992

Josephy, Alvin M. Jr., *The Nez Perce Indians and the Opening Of The Northwest,* Yale University Press, New Haven. 1965.

Kelley. Matt J. *Anaconda, Montana's Copper City,* Soroptimists Club, 1983.

Kitridge, Wm.; Smith Annick, *The Last Best Place.* Montana Historical Society, Helena, 1988.

Klein. Maury, *Union Pacific,* Doubleday, New York, 1987.

Kohrs, Conrad, *An Autobiography,* Gulling Printing, Polson, Montana 1977

Lavender, David, *The Rockies,* Harper & Row, New York, 1975.

Limerick, Patricia N. *Legacy of Conquest,* Norton, New York, 1987.

Meidl, Ruth, Wellcome, George, *Anaconda, Montana A Century of History,* 1983.

Malone, Michael P., *The Battle For Butte,* U. of Wash. Press, 1981.

Malone, Michael P. & Richard B. Roeder, *Montana, A History of Two Centuries,* University of Washington, 1976.

Marcosson, Isaac F. *Anaconda,* Dodd, Meade & Co., New York, 1957.

Morison, Samuel Eliot; Commager, Henry Steele, *Growth of the American Republic,* Oxford Univ. Press, New York, 1942.

McCaffrey, Lawrence J., *The Irish Diaspora in America,* Indiana University Press, Bloomington, 1976.

Oakley, Robert D., *The Philipsburg Story,* U.S. Forest Service, Butte, Montana, 1981.

Prodgers, Jeanette, *Butte-Anaconda Almanac,* Greenfield Printers, Butte, Montana, 1991.

Renshaw, Patrick, *The Wobblies,* Doubleday & Co. Inc. Garden City, New York, 1967.

Rodman, W. Paul, *Mining Frontiers of the Far West,* Holt, Rinehart and Winston, New York, 1963.

Serrin, William, *Homestead,* Random House, New York, 1992

Shoebotham, H. Minar, *Anaconda, Life of Marcus Daly the Copper King,* The Stackpole Company, Harrisburg, Pa.,1956

Spense, Clark C., Montana, *A Bicentennial History,* W.W. Norton & Co. Inc. New York, 1978.

_____, *Territorial Politics and Government in Montana 1864-1869,* University of Illinois, 1975.

Sung, Betty Lee, *The Story of Chinese in America,* Colliers Books, New York, 1967.

Toole, K Ross, *Montana, An Uncommon Land,* University of Oklahoma, Norman Oklahoma, 1959.

Swanberg, W.A., *Citizen Hearst,* Bantam Books, New York, 1971.

Vine, Bob, *Anaconda Memories, 1883 - 1983,* Artcraft Printers, Butte, Montana, 1983.

Waldron, Ellis and Wilson, Paul B., *Atlas of Montana Elections, 1889-1976,* University of Montana, Missoula, 1978.

Wallace, Robert, *The Miners,* Time/Life, 1967.

Writers' Program Montana WPA, *Copper Camp,* Hastings House Publishers, New York, 1943.

Magazines:

Montana, The Magazine of Western History:
    Kelly, Matt J., "Trail Herds of the Big Hole Basin," 57-63. July, 1952.
    Flaherty, Stacy A. "Boycott in Butte: Organized Labor and the Chinese Community 1896-1897", 34-47. Winter 1987.
    Johnson, Carrie, "Electrical Power, Copper and John D. Ryan", 24-37, Autumn, 1988.
    Mercier, Laurie K.. "We Are Women Irish, Gender, Class, Religion, And Ethnic Identity in Anaconda, Montana," 28-41. Winter, 1994.
    Swartout, Robert R. Jr., "From Kwangtun to the Big Sky: The Chinese Experience in Frontier Montana," 42-53, Winter, 1988.

McClure's Magazine:

    Garland Hamlin, "Homestead And Its Perilous Trades," 3-20. June, 1894.

Reports:

    ACM Booklet, "A Brief Description of the Anaconda Reduction Works," 1919, 1926, 1936.
    Bowden, W.E. "The Anaconda Smelter Early Days," Speech 1977, Anaconda Deer Lodge County Historical Society file.

    Knight, Arthur C. MD. "Fifty Years of Progress, report on Montana State Tuberculosis Sanitarium, Galen, Montana," 1963.
    Kramlick, Emil, "ACM Report 1972." Anaconda Deer Lodge County Historical Society file.
    Booklet, "Deer Lodge County 1908." Anaconda Deer Lodge County Historical Society file.
    Waldrup, Hal, Manuscript "The Teepee of the Deer", 1993.

Newspapers:

    Anaconda Standard, January through December, 1889, 1890, 1891, 1892, 1893, 1906, 1907, 1908, 1914, 1917, 1918.

Directories:

    Anaconda City Directory, R.L. POLK & CO'S, 1896, 1898, 1899, 1900, 1902 1905, 1912. 1915, 1916, 1918.
    Butte City Directory, R.L. POLK & CO'S Anaconda Section 1885-86.

INDEX

Dillon, Stanley 37, 55
Dixon, Joseph 250
Dixon, William Wirt 125
Dodge, Gen. Grenville 55
dog racing 89
Dolan, Mary 142
Donnel, R.W. 12
Dougherty, ____ (superintendent at ACM) 223
Douthat, Jim 34
Dragstedt, Rob 111
Driscoll, ____ (sheriff) 248
Dublin Gulch (Montana) 84, 123
Dugan, Larry 294
Duncan, Lewis J. (mayor) 213, 243, 245, 247-248
Dunn, ____ (labor organizer) 266
Dunn, J.P. 107
Dupont Powder Co. 97
Durant, William 233
Durston, Dr. John H. 78-79, 82-83, 107, 141, 157
Duryea brothers 233
Dwyer, Daniel "Dan" (mayor) 30, 34, 118, 157

Eads, William 10
Eardley, John R. 16,17
Eastern Europeans [see immigrants, countries] 101
Eccleston, Henry Harrison 22
Eddy, R.A. 30, 75-76
Edgarton, Sidney 72-73
Edison, Thomas 26, 233
Edwards, Jim 116
Edwards, W.H. "Billy" 66-67
Eggleston, Charles 82, 128, 153, 220
electric trolly 104-05
electrification 227, 229-30, 232
Eliason family 16
Elie, Joseph 111
Elliott, Amos 70
Elliston (Montana) 130
Emeralds Gaelic football team 93
Emma mine 182
Emmons, Bob 92

Emmons, David 76
Emmons, Sam 92
Eneas, Chief 52
English [see also immigrants] 129, 223
Episcopalians 45
Estelle, Minnie 197
Estes & Connell store 80, 110
ethnic integration 291-92
Evans, Gwenilian 16
Evans, Margaret 142
Evans, Miss Lizzie 42
Evans, Morgan 16, 29, 34, 61, 67, 84, 157
Evans Hall (= Evans Opera House) 42, 46, 100, 121, 142, 144-46, 152, 195

Fair Trial Law of 1903 177
Falk, Ben 108, 289
Farlin, William L. 6
Fenner, Robert 34, 141
Ferris, Warren 18,19
fertilizer 239
Fifer, M. 16
Fifth Infantry, U.S. Army 69
Financial Panic (1893) 140, 191
Financial Panic (1907) 221
Finley, Robert 29
Finns 243-244, 248, 253, 261
fire hydrants 48
First Co. Coastal Artillery, U.S. Army 254
First Montana Regiment (Spanish American War) 155
First Presbyterian Church (Anaconda) 263
Fitzgerald, F.D. 47
Fitzgerald, J.D. 80
Fitzpatrick, Barney 92
Fitzpatrick, M.J. 144
Fitzpatrick, Tom 8
Fitzsimmons, Bob (boxer) 88
Flathead county 159, 167
Flathead (Salish) Indians 8, 22, 24, 52, 61-62, 81

Order:
ACONDA MONTANA, COPPER SMELTING BOOMTOWN ON THE WESTERN
NTIER, by Patrick F. Morris. $16.95 plus postage and
dling, $3.50 for first book, 50 cents each additional.

d check or money order to:

SWANN PUBLISHING
5005 BALTAN RD.
BETHESDA, MD 20816

ase send _____ copies of above book to:

E:_____

RESS:_____

Y:_____STATE:_____ZIP:_____

_____

- -

Order:
ONDA MONTANA, COPPER SMELTING BOOMTOWN ON THE WESTERN
NTIER, by Patrick F. Morris. $16.95 plus postage and
lling, $3.50 for first book, 50 cents each additional.

check or money order to:

SWANN PUBLISHING
5005 BALTAN RD.
BETHESDA, MD 20816

se send _____ copies of above book to:

:_____

ESS:_____

:_____STATE:_____ZIP:_____

_____